NASA's First A

Aeronautics from 1958 to 2008

National Aeronautics and Space Administration
Office of Communications
Public Outreach Division
History Program Office
Washington, DC
2013

The NASA History Series
NASA SP-2012-4412

NASA's First A

Aeronautics from 1958 to 2008

Robert G. Ferguson

Library of Congress Cataloging-in-Publication Data

Ferguson, Robert G.
NASA's first A : aeronautics from 1958 to 2008 / Robert G. Ferguson.
p. cm. -- (The NASA history series) (NASA SP ; 2012-4412) 1.
United States. National Aeronautics and Space Administration--History. 2.
United States. National Advisory Committee for Aeronautics--History. I.
Title.
TL521.312.F47 2012
629.130973--dc23

2011029949

This publication is available as a free download at ***http://www.nasa.gov/ebooks.***

ISBN 978-1-62683-009-7
90000>
9 781626 830097

TABLE OF CONTENTS

ACKNOWLEDGMENTS

Before naming individuals, I must express my gratitude to those who have labored, and continue to do so, to preserve and share NASA's history.

I came to this project after years of studying private industry, where sources are rare and often inaccessible. By contrast, NASA's History Program Office and its peers at the laboratories have been toiling for five decades, archiving, cataloging, interviewing, supporting research, and underwriting authors. For a survey work such as this, the ability to quickly locate primary documents and published historical accounts has been a necessity and a pleasure. I am deeply indebted to countless individuals, some of whom I know, many of whom I do not, who have quietly saved documents, patiently cared for photographs, and generously shared these treasures.

I wish to thank the NASA scientists and engineers who eagerly told their stories—I hope that I have done their stories justice. I also wish to thank the many researchers who took the time to cast their work within a historical context.

I need to make special mention of the historians who have gone before me and illuminated many of the case histories upon which this book depends. This undertaking would have been dramatically different, and probably impossible in my lifetime, had it been necessary to investigate every significant research program from scratch. While I share a general concern about the use of secondary sources, the authors upon whom I have relied have been exceedingly thorough and skilled.

Many at NASA provided their expert assistance, including Jane Odom, Colin Fries, John Hargenrader, Kathryn Goldberg, Anthony Springer, G. Michael Green, Glenn Bugos, Gail Langevin, and Nadine Andreassen. I also am indebted to Steve Garber, Glen Asner, Steve Dick, and Bill Barry for their patience, counsel, and encouragement.

I also wish to thank my reviewers, who waded through a difficult manuscript. For introducing me to this particular topic and assisting me at the beginning of my research, I thank both Roger Launius and Erik Conway.

A number of talented professionals helped transform my manuscript into a finished book. Kay Forrest carefully copyedited the manuscript. In the NASA Headquarters Communications Support Services Center, Lisa Jirousek, George

Gonzalez, and Kurt von Tish performed additional copyediting; Christopher Yates designed and laid out the book; and Tun Hla oversaw the printing process.

I have gained much over the years from my network of colleagues in the history of technology. Members of the Society for the History of Technology, especially those aerospace historians known affectionately as "the Albatrosses," were extremely helpful in a variety of large and small ways.

Lastly, I wish to thank my wife and children and all those I have not acknowledged for this opportunity to explore, reflect, and tell a story.

Chapter 1:

The First A: The Other NASA

That NASA is more than a space agency may come as a surprise to some. Aeronautics, the first *A* of the NASA acronym, has always been a part of the National Aeronautics and Space Administration, but against the headline exploits of rocket launches, Moon landings, Space Shuttle missions, and Mars rovers, aeronautics is easily lost in the shadows of NASA's marquee space programs. This relative obscurity belies what has been a remarkably creative, productive, and highly effective group of researchers who, at one time, even helped bring about the Space Age and invent a space agency. The list of accomplishments for NASA's first *A* is long, and this book goes a modest way toward sketching these developments.

Aeronautics really might be called the "other NASA," distinct in its charge, methodologies, and scale. Aeronautics research is not mission-oriented in the same way that going to the Moon or Mars is. It is interested in learning about physical phenomena, such as turbulence, and *how* to do something, such as quieting the noise of helicopter blades. Aeronautics' mission is not about going somewhere specific or building one particular thing; it is about supporting the country's commercial and military needs with respect to aviation. The contrast with the space program is telling: where NASA has gone to considerable lengths to justify the space program and explain how space innovations feed into the country's more terrestrial needs, aeronautics research has had a direct and undeniable impact on commercial and military technology development. Its positive economic effects are generally accepted.[1]

1. A summary of studies on NASA's economic impact may be found in U.S. Congress, Congressional Budget Office (CBO), "Reinventing NASA," March 1994, pp. 2–4. Relating specifically to the question of government-funded aeronautical research, see David C. Mowery and Nathan Rosenberg, *Technology and the Pursuit of Economic Growth* (New York: Cambridge University Press, 1989), pp. 169–202; and George Eberstadt, "Government Support of the Large Commercial Aircraft Industries of Japan, Europe, and the United States," chap. 8 in *Competing Economies: America, Europe, and the Pacific Rim*, OTA-ITE-498, by U.S. Congress,

Being research-oriented, aeronautics naturally has a different methodological bent. Its activities revolve almost entirely around experimentation, made up of a combination of cutting-edge basic research and more routine assistance to its patrons in the military and industry. It has encompassed the use of wind tunnels,[2] the flight testing of revolutionary prototypes, and the creation of new methodologies, such as computational fluid dynamics. Finally, aeronautics research typically has been conducted on a small scale and on, relatively speaking, limited budgets. It has not been characterized by extremely large projects on the order of the Apollo or Space Shuttle programs. This is not for lack of desire within the aeronautical research community; rather, the political will to do in aeronautics what was done for the space program has not been similar. Ultimately, the bulk of aeronautics work has been performed on a fraction of NASA's overall budget. Even so, NASA's aeronautics has had an outsize impact.

The nature of this impact deserves special attention because it locates NASA's work within larger technical and historical narratives. The risk of detailing NASA's technical achievements is to ascribe a kind of primacy and distinctiveness that is characteristic only some of the time. NASA researchers were not working in ivory towers, and they were not always the first link in a chain of innovation stretching from laboratory to industry. It is important to appreciate the ebb and flow of scientific and technological knowledge into and out of the various communities represented by NASA's first *A*. NASA was not alone in pursuing aeronautical research: the U.S. military had its own facilities and programs, as did industry, academia, and foreign governments. Published papers, conferences, technical committee meetings, and joint projects created a shifting area of known and unknown, possible and impossible, across all of these actors. Researchers picked up where others left off, often taking ideas in new directions and, in a spirit of collegiality, hoping to best the others. This competition extended to the various communities within NASA as well.

That said, NASA has occupied a unique place within the research landscape. It has been a pooled national investment, providing a cadre of talented

Office of Technology Assessment (Washington, DC: U.S. Government Printing Office [GPO], October 1991).

2. Wind tunnels, to be brief, are laboratory devices for measuring the flow of air, especially as it moves around solid shapes. They may be used to simulate the performance of aircraft and parts of aircraft in a highly controlled environment, thus reducing the need for risky and potentially costly flight testing. There are plenty of aeronautical investigations that do not need or make use of wind tunnels. It is but one experimental device available, and it finds its greatest utility in such areas as aerodynamics (e.g., lift and drag), aeroelasticity, thermodynamics, control, noise, propulsion, and icing.

researchers as well as facilities beyond industry's means. Its collection of experimental facilities and flight-test capabilities was and remains unparalleled in the United States.[3] Since World War II (WWII), it finds international comparison only to Russia's Central Aerohydrodynamic Institute (TsAGI) and Gromov Flight Research Institute (both in Zhukovsky) and, more recently, the loosely associated but well-supported European experimental facilities.[4] Along with such advanced experimental capabilities, NASA's researchers had the institutional space to pursue more basic, more long-range, and riskier projects than many of their non-NASA peers. Though NASA has had its share of routine aeronautical testing, such as debugging problems for industrial partners, the organization gave researchers the freedom to pursue questions without commercial justification or market deadlines and without the narrow confines of military applicability. NASA researchers saw their positions as dream jobs, not for the money (they likely could have earned more in industry), nor for the prestige (none became household names), but for the ability to push the boundaries of what was known and what was technically possible. Such ideal research conditions drew talented engineers and scientists who, armed with some of the most capable and exotic experimental equipment and charged with the job of ensuring U.S. preeminence in military and commercial aviation, returned a panoply of remarkable, sometimes revolutionary, discoveries and technologies. These contributions have largely been invisible to the public.

Step onto a modern commercial passenger jet and you will be surrounded by some of the most indispensable of NASA's contributions. The wings, for example, are likely descendants of the supercritical airfoils first developed

3. NASA's aeronautics infrastructure was estimated to have a replacement value of $10 billion in 1984. See Eberstadt, "Government Support of the Large Commercial Aircraft Industries of Japan, Europe, and the United States," p. 347.
4. TsAGI is more properly the Tsentralniy Aerogidrodinamicheskiy Institut. The major experimental agencies and facilities in Western Europe include the following: in France, the Office National d'Études et de Recherches Aérospatiales (ONERA); in the United Kingdom, QinetiQ and the Defence Science and Technology Laboratory; in Germany, the Deutsches Zentrum für Luft- und Raumfahrt (DLR) and the European Transonic Wind Tunnel; and in the Netherlands, the Nationaal Lucht- en Ruimtevaartlaboratorium (NLR). France, Germany, and the Netherlands have, in recent years, sought to more closely integrate their research infrastructure. There are additional laboratories in Belgium, Switzerland, Italy, Sweden, and the Czech Republic. See Library of Congress, Federal Research Division, "Aeronautical Wind Tunnels in Europe and Asia" (Washington, DC, February 2006); and Philip S. Antón et al., "Wind Tunnel and Propulsion Test Facilities: Supporting Analyses to an Assessment of NASA's Capabilities to Serve National Needs" (RAND Corporation Technical Report 134, 2004).

at the Langley Research Center in the 1960s and demonstrated in the early 1970s. The control system, if it is of the latest variety, is a digital fly-by-wire system, a technology developed and first demonstrated at the Dryden Flight Research Center in the early 1970s. The aircraft itself, if designed in the last two to three decades, benefited from computational fluid dynamics (CFD), a way of simulating the aerodynamic flows about the aircraft using computers. NASA was a pathbreaker for CFD, aggressively pursuing CFD as both a potential replacement for and a complement to wind tunnel studies. Beyond commercial aircraft, aeronautical engineers created and tested novel reentry systems for returning space vehicles. They helped develop tilt-rotors, vehicles that combine the benefits of helicopters and fixed-wing aircraft. As of the writing of this book, they are helping to renovate the nation's air traffic control system. More fundamentally, they have run thousands and thousands of tests on all manner of aviation-related questions, such as acoustics, aerodynamic drag, icing, vibration, crash survivability, and engine efficiency, just to name a few. These are the truly invisible contributions, evidenced by laboratory logs, conference presentations, technical reports, and journal articles. These reports and articles have been read, some filed away, some debated, some expanded and reformulated. Some sparked new solutions, and some spurred designers to try new approaches. Such has been NASA's place in the larger ecology of aeronautical knowledge.

Still, one may wonder why NASA is involved in aeronautical research in the first place, and why in this manner? Federal support dates not merely to the creation of NASA, but to 1915 and the creation of NASA's predecessor, the National Advisory Committee for Aeronautics (NACA, pronounced letter by letter).[5] The Wright brothers first flew in December 1903, but only with the use of aircraft in World War I did aviation became a matter of national importance. This technology and the passions and fears it provoked brought about the NACA, a civilian government agency devoted entirely to the military and commercial development of a single technology. Certainly this was not the country's first government laboratory, and it was not the first attempt to provide research for the country's economic well-being. The Army and Navy's testing facilities, the Naval Observatory, and the National Bureau of Standards

5. Sticklers will note that the U.S. government supported aeronautical activities as early as the Civil War with military ballooning and, at the turn of the century, through Samuel P. Langley's aerodrome experiments conducted under the aegis of the Smithsonian. The NACA, however, was the first ongoing laboratory. Tom D. Crouch, *Wings: A History of Aviation from Kites to the Space Age* (New York: W. W. Norton, 2003), pp. 46–48, 76. F. Stansbury Haydon, *Military Ballooning During the Early Civil War* (Baltimore: Johns Hopkins University Press, 2000).

all predated the NACA, as did the country's agricultural experiment stations begun under the Hatch Act in 1897 and the land grant colleges initiated under the Morrill Act of 1862,[6] but there was no national laboratory for automobiles or railroads. Such technologies were the province of private industry.[7] Aircraft evoked a different response. Here, a group of individuals from military, academic, and government backgrounds managed to exploit fears about America's technological preparedness to create an aviation laboratory. European advances in aviation, as well as strong national support for aircraft development in England, France, and Germany, played no small part in encouraging the U.S. government to act.[8]

There were alternatives to the NACA's civilian laboratory. The government could have opted for more narrowly focused military aeronautics laboratories, not unlike what the U.S. Air Force created after World War II. The government also could have supported university-run experimental facilities, something that the Guggenheim Fund for the Promotion of Aeronautics achieved privately in the interwar years.[9] The NACA's government-owned

---

6. A. Hunter Dupree, *Science in the Federal Government: A History of Policies and Activities* (Baltimore: Johns Hopkins University Press, 1986). The Army began conducting simple laboratory work at the Watertown Arsenal in Massachusetts as early as 1821; see Libby Baylies Burns and Betsy Bahr, "Watertown Arsenal, Watertown, Middlesex County, MA," Historic American Engineering Record report no. MA-20 (Washington, DC: National Park Service, 1985), pp. 136–137. The Army's first proving ground was the Sandy Hook Proving Ground in New Jersey, established in 1874. The U.S. Navy's Washington Navy Yard performed testing in the 19th century and installed a towing tank in 1898. See Taylor Peck, *Round Shot to Rockets: A History of the Washington Navy Yard and the Naval Gun Factory* (Annapolis: Naval Institute Press, 1949); Steven J. Dick, *Sky and Ocean Joined: The U.S. Naval Observatory, 1830–2000* (Cambridge, U.K.: Cambridge University Press, 2003); and A. C. True, "Agricultural Experiment Stations in the United States," *Yearbook of the United States Department of Agriculture, 1899* (Washington, DC: GPO, 1900), pp. 513–548.
7. The Department of Transportation eventually created a High Speed Ground Transportation Test Center (for rail technology) near Pueblo, CO, in the early 1970s. In 1982, the Association of American Railroads took over the operation and renamed it the Transportation Technology Center. The Center remained one of the principal laboratories for federally funded railroad research. U.S. Department of Transportation, Federal Railroad Administration, "Five-Year Strategic Plan for Railroad Research, Development, and Demonstrations," March 2002.
8. Alex Roland, *Model Research: The National Advisory Committee for Aeronautics, 1915–1958* (Washington, DC: NASA SP-4103, 1985).
9. Reginald M. Cleveland, *America Fledges Wings: The History of the Daniel Guggenheim Fund for the Promotion of Aeronautics* (New York: Pitman, 1942); Richard P. Hallion, *Legacy of Flight: The Guggenheim Contribution to American Aviation* (Seattle: University of Washington, 1977).

and -operated laboratory provided something of both: an institution that could perform military research (and guard such closely held information) as well as more academic and generally useful experimental studies. As a civilian laboratory, it could address the dual concerns of military preparedness and commercial competitiveness.

It is conceivable that, in the absence of government support, industry might have eventually created its own pooled research infrastructure. During World War II, a number of companies did band together to build a cooperative wind tunnel; the arrangement and the tunnel, however, did not have a long lifespan.[10] There have also been private actors, such as Calspan (the descendent of Curtiss-Wright's Buffalo labs and the Cornell Aeronautical Laboratory), that have provided for-hire research and development services.[11] Also, companies supported, early on, parallel research and teaching programs at nearby colleges and universities (e.g., Boeing at the University of Washington, Douglas Aircraft at the California Institute of Technology [Caltech]).[12] Thus, templates for the privatization of research have existed, but this path has never been fully explored and remains an unlikely scenario. In light of government assistance in Europe and the former Soviet Union (depending on when the comparison is made) and an overriding and persistent national desire to stay at the forefront of this particular technology, it has been a foregone conclusion that the U.S. government would also underwrite much of the country's own basic research. This global race is very much about competing models of government support for, and intervention in, a particular industry (more broadly defined as the aerospace sector).

There have been at least two additional factors that have buttressed the economic and military arguments in favor of federally funded research. The first is the place that aeronautics has as a cultural force and national symbol. The emergence of the airplane was a fantastic and awesome spectacle. The

10. Robert G. Ferguson, "Technology and Cooperation in American Aircraft Manufacture During World War II" (doctoral diss., University of Minnesota, 1996).
11. Kevin R. Burns, "The History of Aerospace Research at Cornell Aeronautical Laboratory" (American Institute of Aeronautics and Astronautics [AIAA] Space 2004 Conference, San Diego, CA, 28–30 September 2004 [AIAA publication 2004-5884]).
12. J. Lee, D. S. Eberhardt, R. E. Breidenthal, and A. P. Bruckner, "A History of the University of Washington Department of Aeronautics and Astronautics, 1917–2003," in *Aerospace Engineering Education During the First Century of Flight*, by Barnes W. McCormick, Conrad F. Newberry, and Eric J. Jumper, AIAA Library of Flight Series (Reston, VA: AIAA, 2004); Arthur E. Raymond, interview by Ruth Powell, Pasadena, CA, 2 April 1982, transcript, California Institute of Technology Oral History Project, Caltech Archives.

first public flights moved quite a few people to decide that aviation was their passion and, if possible, their vocation. Aviation was heroic and modern. It represented the pinnacle of technological development.[13] Though spaceflight has eclipsed aeronautics in this respect, aviation remains symbolic of national achievement and human ambition. It is no accident that of all the technologies that might be given a place of honor in the nation's capital, only aviation and spaceflight have their own dedicated museum on the National Mall.[14] The nation's commitment to aeronautical prowess is, in part, about retaining a technological crown.[15]

The second contributing factor concerns the pace of technological change. An aeronautical laboratory only makes sense in the context of an evolving and little-understood technology. Had aviation somehow matured in a decade or two and had all the underlying physical phenomena revealed their secrets, there would have been little reason to continue further experimentation.[16] Instead, the last century of flight has seen a series of developments that have fed calls for more research. Engineers pushed aircraft performance to faster speeds and higher altitudes, and, when they reached practical and economic barriers to going faster and higher, they turned their attention to efficiency, safety, and maneuverability. It is only more recently, some have argued, that the state of aeronautical knowledge has reached levels of maturity that portend quieter days for some of the laboratories. Still, aeronautical research remains sufficiently fecund that as larger tectonic shifts occur in the global economy, politics, the environment, and society, our definition of the technological frontier will also shift, leaving the laboratories with new puzzles to solve.

---

13. Joseph J. Corn, *The Winged Gospel: America's Romance with Aviation* (Baltimore: Johns Hopkins University Press, 2002); Bayla Singer, *Like Sex with Gods: An Unorthodox History of Flying*, Centennial of Flight Series (College Station: Texas A&M Press, 2003).
14. Automobiles and trains, arguably of greater social and economic importance than aircraft, can be found on the National Mall, but only within the more general collections of the National Museum of American History.
15. The United States is not unique in this respect as quite a few countries trumpet aviation (and space) achievements as elements of their national ethos. See, for example, Scott W. Palmer, *Dictatorship of the Air: Aviation Culture and the Fate of Modern Russia* (Cambridge, U.K.: Cambridge University Press, 2006); and Peter Fritzsche, *A Nation of Fliers: German Aviation and the Popular Imagination* (Cambridge, MA: Harvard University Press, 1992).
16. What constitutes a mature technology is an easy source of academic disagreement. Here, I take maturity to be something that is socially defined. For further exploration of this subject, a fine starting point is Wiebe E. Bijker, *Of Bicycles, Bakelites, and Bulbs: Toward a Theory of Sociotechnical Change* (Cambridge, MA: Massachusetts Institute of Technology [MIT] Press, 1997).

The Technical Narrative

In a classic and controversial essay on the history of technology, economic historian Robert Heilbroner argued that technology unfolds according to a logical sequence, that levels of scientific knowledge and technical skill are controlling factors for what follows next.[17] A similar argument might be made for the unfolding of aeronautical knowledge in the 20th century as discoveries and innovations set the context for subsequent discoveries and innovations. The rise of computational fluid dynamics, for example, paced by the measured and seemingly predictable advance of semiconductors, has had knock-on effects across the organization. The development of digital fly-by-wire, likewise, has been an enabling technology for a whole new class of unstable vehicles that could not fly without such automatic control. Still, that there is an order to technical development obscures the manner in which the various aeronautical actors steered research one way or another; so too, such a conception ignores the manner in which political, military, and economic contexts swayed the paths of research. One can tell the story of NASA as a march of progress, but it is more honest and more interesting to understand how individual initiative and historical context wove the technical narrative.

The narrative about NASA's aeronautics begins not in 1958, but in 1915, with the creation of the NACA. For the next quarter century, the NACA's Langley Laboratory would establish a workmanlike reputation for its aerodynamic investigations, providing a fledgling industry and military aviation with key design data and insights.[18] World War II and its aftermath greatly altered the landscape for aviation, catapulting the role of military aviation beyond its prewar, adjunct status. Military aviation proponents gained a new service branch, the Air Force, and a Navy with aircraft carriers, not battleships, at the heart of their battle groups (i.e., task forces and task groups).[19] The aircraft industry found itself on a new political and financial footing, no longer struggling from one small batch of aircraft to the next as it had before the war. Commercial aviation, likewise, was growing and profitable.

17. Robert L. Heilbroner, "Do Machines Make History?" *Technology and Culture* (8 July 1967): 335–345.

18. Roland, *Model Research*; James R. Hansen, *Engineer in Charge: A History of the Langley Aeronautical Laboratory, 1917–1958* (Washington, DC: NASA SP-4305, 1987).

19. Herman S. Wolk, *Planning and Organizing the Postwar Air Force, 1943–1947* (Washington, DC: Office of Air Force History, 1984); Norman Friedman, *The Postwar Naval Revolution* (Annapolis: Naval Institute Press, 1986); Jeffrey G. Barlow, *Revolt of the Admirals: The Fight for Naval Aviation, 1945–1950* (Washington, DC: U.S. Naval Historical Center, 1994).

The industry now had two legs to stand on.[20] The NACA, too, had blossomed into three Laboratories with two specialized flight-test areas, one for aircraft and one for rocketry. The rapid ascension of the technology and its military and commercial acceptance generally augured well for the NACA, but success gave rise to additional actors. Not only did industry greatly expand its laboratory infrastructure, but the U.S. Air Force had also begun to create its own in-house and subcontracted research and development (R&D) capabilities. All three of the military arms ultimately pursued their own missile programs. As to matters of leadership and the nominal task of serving as the nation's aeronautical advisor, the NACA found itself treading gingerly between powerful and vocal partners.[21]

From 1945 to 1957, the dominant aeronautical themes revolved around speed and altitude. Jet engines and rockets were enabling technologies in this regard, but the research direction was also a complement to the Cold War and the spread of atomic weapons. Wartime lessons about R&D and technological leadership stoked fears that the Soviet Union might attain and wield superior military capabilities. High-speed, high-altitude flight was one such Cold War battleground, for these attributes, at least according to military aviators, tipped both offensive and defensive scales. Technically, this direction called for a whole new class of research facilities and methodologies. Of particular difficulty was predicting aerodynamic behavior in the transonic region, that is, the region straddling Mach 1, or the speed of sound.[22] To

20. The industry did suffer an immediate postwar downturn as the government canceled wartime contracts, but over the long term, the industry experienced a level of support and prosperity that had eluded it prior to WWII. Donald M. Pattillo, *Pushing the Envelope: The American Aircraft Industry* (Ann Arbor: University of Michigan Press, 1998), chaps. 9–11; Carl Solberg, *Conquest of the Skies: A History of Commercial Aviation in America* (Boston: Little, Brown, 1979); R. E. G. Davies, *Airlines of the United States Since 1914* (Washington, DC: Smithsonian Institution Press, 1982).
21. Jacob Neufeld, *The Development of Ballistic Missiles in the United States Air Force, 1945–1960* (Washington, DC: Office of Air Force History, 1990); Graham Spinardi, *From Polaris to Trident: The Development of U.S. Fleet Ballistic Missile Technology* (Cambridge, U.K.: Cambridge University Press, 1994); John W. Bullard, *History of the Redstone Missile System*, U.S. Army Missile Command Historical Monograph AMC-23-M (Redstone Arsenal, AL: U.S. Army, October 1965); Michael J. Neufeld, *Von Braun: Dreamer of Space, Engineer of War* (New York: Alfred A. Knopf, 2007).
22. Mach, named for Austrian physicist Ernst Mach, is a dimensionless measure of an object relative to the speed of sound. Mach 1 means that something is traveling at the speed of sound; Mach 1.5 means that it is traveling at one and a half times the speed of sound. The actual speed of

sidestep the difficulties encountered when wind tunnels attempted to model the transonic regime, the NACA and the military set out to create a series of experimental vehicles, the first and most famous being the Bell X-1, which rocketed Chuck Yeager past Mach 1. Though researchers at the Langley Laboratory eventually discovered how to ingeniously design wind tunnels for the transonic region, the move into X-planes established a new and long-lasting community of flight-test-oriented researchers. This was a fortuitous turn; instead of flight testing serving as a kind of ancillary laboratory function, the community of engineers, pilots, and technicians working on the dry lakebeds of Muroc, California, turned flight testing into a creative activity, a starting point for new and fruitful investigations.[23] Geopolitics *and* thorny physical phenomena nurtured research methodology.

Sputnik's October 1957 launch and orbit upended the NACA. A year later, in the NACA's place, the country created a civilian space and aeronautics agency. To the original NACA Laboratories, Congress and President Dwight D. Eisenhower added the various military rocketry programs and the charge of answering Soviet space initiatives. For a time, the nation's space program operated out of Langley Research Center, led by many former NACA aeronautics researchers. Most aeronautics programs initially continued to operate as they had before the change to NASA, but these were now a minor portion of a much larger and qualitatively different enterprise. Representing the acme of the higher and faster era of X-planes, NASA and the Air Force conducted rocket-plane flights at the edge of space in the X-15. Reaching speeds above Mach 6 and altitudes 67 miles above Earth, the X-15 was as much spacecraft as aircraft. Another experimental vehicle, begun through the personal initiative of an engineer at Muroc, paved the way for the creation of spacecraft that could fly to a landing when returning to Earth. The lifting-body program, as it was known, captured both the enterprising nature of grassroots projects and a genuine interest among aeronautical engineers to contribute to the space program.[24]

---

sound through a gas varies depending on the medium through which the sound travels and its temperature; in the case of aircraft, this means that sound travels faster at lower altitudes where the air is warmer. See John D. Anderson, Jr., *A History of Aerodynamics and Its Impact on Flying Machines* (Cambridge, U.K.: Cambridge University Press, 1998), pp. 375–376.

23. Richard P. Hallion, *On the Frontier: Flight Research at Dryden, 1946–1981* (Washington, DC: NASA SP-4303, 1984); Hansen, *Engineer in Charge*, chaps. 9–11.

24. R. Dale Reed, with Darlene Lister, *Wingless Flight: The Lifting Body Story* (Washington, DC: NASA SP-4220, 1997); Hallion, *On the Frontier*.

One revealing project of the 1960s that NASA did not oversee was the Supersonic Transport, or SST. NASA served as the research arm for this Federal Aviation Administration (FAA) project. Had NASA's aeronautics been equal in importance to the mission-oriented space side of the Agency, this might well have been a full-fledged NASA-managed project. However, the Agency had its hands full with its primary focus, going to the Moon. What is also telling is the fact that the nation pursued, however briefly, the SST project at all. The U.S. government, though supportive of aeronautics, had never crossed the line to underwriting directly the development of a commercial aircraft.[25] The aircraft had its boosters (in industry, at the FAA, and at NASA) who supported the technology because it was the "next step," but what gave it special impetus was the threat posed by Europe's government-subsidized effort, Concorde. Even though Concorde ultimately proved unable to deliver on its commercial promises, it was the beginning of a unified and resurgent European aeronautical industry. Over the next five decades, NASA and its supporters would point to the growing European competitive threat (and its underlying government subsidies) as justification for U.S. support of aeronautical research. Still, and in spite of proffers to use NASA to develop exceedingly advanced commercial aircraft, NASA's contribution to the economy continued to reflect a longstanding political philosophy of limited industrial intervention.

One of the more significant developments of the 1960s was one of the least flashy. Eschewing the risk and prestige of supersonic aircraft design, Richard Whitcomb and his team at Langley explored how to design more efficient subsonic airfoils and wings. Their solution, validated using airfoil models in wind tunnels, provided a template for all future low-drag, high-subsonic-speed wings.[26] Whitcomb's work was unusually insightful, and it exemplified the quiet and persistent efforts to understand and improve aircraft performance. It was also representative of the engineer's art, where answers are rarely found in singular equations or inventions, but in the complex balancing of competing demands and physical phenomena. Whitcomb would go on to design and test winglets in the 1970s, establishing them as a useful option for engineers seeking greater subsonic efficiency.

---

25. The U.S. government's boldest attempt to support commercial aviation was the subsidization of airmail in the interwar years, a practice that greatly assisted both the airlines and manufacturers. Henry Ladd Smith, *Airways: The History of Commercial Aviation in the United States* (Washington, DC: Smithsonian Institution Press, 1991), chaps. 7–8; Crouch, *Wings*, chaps. 6–7.

26. John D. Anderson, Jr., *A History of Aerodynamics and Its Impact on Flying Machines* (Cambridge, U.K.: Cambridge University Press, 1998), pp. 413–417.

The decade of the 1970s was a turning point for the technical, political, and economic threads of this narrative. The drawdown of the Apollo program (the final Moon mission took place in December 1972) brought the Agency into a kind of existential crisis, and it came at a time of environmental protest, technological backlash, and energy shortages. Highlighting aeronautics research was one way that NASA could show it was relevant to everyday problems. Answering public concern over oil shortages, NASA created the Aircraft Energy Efficiency Program, a multipronged effort that sought to provide a knowledge base for more fuel-efficient commercial aircraft. Responding to growing air traffic congestion, NASA highlighted its work on short takeoff and landing aircraft for possible commercial use. Though these research programs had the appearance of being politically generated (i.e., following political demands of the President, Congress, and constituents), they were, in fact, existing lines of research. Yet this shift in the packaging of aeronautics research was hardly symbolic; it presaged an increased level of control over research projects by Headquarters.

The 1970s were the breakout decade for digital computing. NASA had long used computers for performing tedious mathematical calculations, and in the 1960s the Langley Laboratory oversaw the development of a structural analysis software package (NASTRAN, or NASA Structural Analysis) that would become widely used in the engineering world; but with the 1970s, NASA put digital computing to new use. At Ames Research Center, some researchers aggressively set out to replace wind tunnels through the use of supercomputers. At Dryden, engineers created a digital fly-by-wire control system that, to put it simply, decided how best to fly the aircraft, and at Langley, engineers worked to integrate new digital equipment into the cockpit. In hindsight, the spread of computers might appear entirely logical and expected, but what researchers were setting out to do was revolutionary in every way. Digital fly-by-wire, for example, was not simply about replacing manual control with computer control, but about designing aircraft that humans could not control without computers. Digital computing helped redefine practice in the laboratory, in the factory, and in the air.

The attention paid to aeronautics in the post-Apollo era (along with tighter Agency funding) brought about a shuffling of Center specialization. At Lewis, which had focused almost exclusively on space propulsion in the 1960s, the Center returned to its roots and began experimenting with air-breathing propulsion again. At Ames and Langley, two Centers that appeared (to some) to duplicate each other's research capabilities, concern grew about how to differentiate the two. In fact, there had always been more than enough research work to go around, but looking down from the top, Ames and Langley both had large expanses of wind tunnels. The outcome was that Ames diversified into areas beyond its aeronautics core, pursued computational flight dynamics as a gambit to leapfrog wind tunnels, and gained control of NASA's vertical and short takeoff programs. This put Langley in charge of the long-haul, fixed-wing programs, as well as structures, materials, guidance and control, and environmental quality.

In the 1980s, a loosening of oil supplies reduced concerns about fuel efficiency while reinvigorated fears of the Soviet Union increased funding for military development and new experimental vehicles. Playing into the Cold War push were changes in battle tactics that emphasized stealth and maneuverability over speed and altitude.[27] NASA's digital fly-by-wire technology proved to be an enabling technology for many of these highly unstable designs. While many of the vehicles of this X-plane renaissance were fruits of the Department of Defense (DOD), NASA's Langley and Dryden facilities provided essential assistance in wind tunnel evaluation and flight testing. The 1980s also saw the introduction of three major test facilities: the creation of the National Transonic Facility at Langley (1982), the addition of a new 80-by-120-foot open section to Ames's National Full-Scale Aerodynamics Complex (1987), and the creation of Ames's Numerical Aerodynamic Simulation facility (1988).

In 1986, NASA's Space Shuttle Challenger exploded during launch. The tragedy forced the Agency into a period of public scrutiny and self-examination. Leaky O-rings on the Shuttle's solid rocket boosters were eventually tied to failures in leadership, program management, and Agency culture.[28] A little over a week after the accident, President Ronald Reagan announced that the country would be moving ahead with the Space Shuttle, a space station, and a new project called (at least in his State of the Union Address) the Orient Express. Though advertised primarily as a commercial passenger vehicle, the National Aero-Space Plane (NASP) was really the second phase of a single-stage-to-orbit project for the military. NASA joined the project and performed research on scramjets, hypersonic aerodynamics, and materials, all areas in which the Agency had prior experience. Indeed, Langley's longtime work on scramjets had helped inspire the military's earlier designs. NASP never realized its goal, but it did

---

27. Some aspects of the historical shift in tactics are captured in Robert Coram, *Boyd: The Fighter Pilot Who Changed the Art of War* (New York: Little Brown and Company, 2002). On stealth development, see David C. Aronstein and Albert C. Piccirillo, *Have Blue and the F-117A: Evolution of the "Stealth Fighter,"* AIAA Library of Flight Series (Reston: AIAA, 1997).
28. Diane Vaughan, *The Challenger Launch Decision: Risky Technology, Culture, and Deviance at NASA* (Chicago: University of Chicago Press, 1996).

answer those critics who said that NASA needed to regain the bold spirit of the Apollo program.

The 1990s signaled a series of turning points in the Agency's aeronautics programs. Externally, the end of the Cold War spelled fewer military research and development programs; it also lessened one of the key rationales for federally funded aeronautics research. At a time when the forthcoming generation of American fighter aircraft (i.e., the eventual F-22 and F-35) was going unanswered by potential adversaries, research on even more advanced technologies was hardly pressing. Further, within the federal government and within NASA, moves were afoot to make the Agency more accountable and businesslike, trends that reduced grassroots control over research and budgetary decisions. Daniel Goldin, who served as NASA Administrator from 1992 to 2001, championed these changes and called on aeronautics to reinvent itself. What followed for the next decade was a series of mission statements and restatements, each with consequent organizational tweaks and upsets. Most immediately, NASA's aeronautics programs attempted to emulate NASA's large-scale programs by situating its research within two massive programs: the Advanced Subsonic Technology (AST) Program and the High Speed Research (HSR) Program. For a time, the strategy appeared successful: NASA's annual aeronautics funding approached 1 billion dollars (or even more, depending on how one measures the aeronautics budget), with AST and HSR funding a broad array of projects, such as engine noise and composite material applications. However, the Agency was never greatly invested in the two megaprograms, so when budgetary pressures became too great, AST and HSR tumbled. Program size proved no bulwark against spending cuts. By 1999, funding for AST and HSR came to an end, and, from a budget high of $957 million in 1994, funding dropped to $600 million in 2000 and continued retreating into the next decade.[29]

By the turn of the millennium, aeronautics was caught in the doldrums, beset by shifting program goals and uncertain support for much of the basic research that had been its mainstay. The loss of the Shuttle Columbia on reentry in 2003 brought about another reexamination of the human space program; this ultimately led to a decision to continue building the International Space Station while also designing and building a new space transportation system that would take humans back to the Moon and on to Mars. All of this was to be accomplished without a dramatic increase in Agency funding. In aeronautics, these more recent events only exacerbated longer-term trends that were whittling away at core programs. Concern grew about the viability of wind tunnels

29. See this volume's appendix.

and expert communities. The rotorcraft program at Ames, for example, almost disappeared, leaving for a period only the Army half of the rotorcraft laboratory. In 2005, a new Administrator, Michael Griffin, reoriented the Agency, this time returning the Aeronautics Mission Directorate's focus to basic research. Whereas the prior decade had sought to buttress aeronautics through a futile search for politically attractive applications and end products, Griffin's policy stated that basic research was the goal. Though a welcome change, the policy did little to stem the overall decline in funding.

Communities of Researchers

Aeronautics was never a singular entity or function within NASA. It comprised activities at four Research Centers. Langley in Virginia, the oldest of the Laboratories, has been the most closely associated with aviation and often characterized as engineering-minded and pragmatically oriented.[30] The Lewis Lab in Ohio, renamed the Glenn Research Center in 1999, focused largely on space propulsion during the 1960s but revitalized its jet and propeller turbine research in the 1970s.[31] Not all propulsion work has been done at Lewis/Glenn; Langley has long been the center of research for scramjets, very specialized engines designed to operate at hypersonic speeds. In California, Ames initially was a complement to Langley and included a large number of similar wind tunnel facilities. Over time, Ames established itself as the more theoretically oriented of the two and worked to diversify itself away from Langley's strengths by developing astrobiology research, CFD, human factors, flight simulation, airspace management, and "short-haul" aircraft (e.g., rotorcraft).[32] Dryden, initially a simple flight research base amidst California's dry lakebeds, blossomed into a very capable and innovative research facility in its own right.[33]

Within each one of these Research Centers were various research branches, teams of scientists and engineers focused on a particular problem area or subject. Early on, the branches at Langley and Ames were typically associated with particular speed regimes and a specific wind tunnel. Other branches

---

30. On the Langley Lab, see Hansen, *Engineer in Charge*; and James R. Hansen, *Spaceflight Revolution: NASA Langley Research Center from Sputnik to Apollo* (Washington, DC: NASA SP-4308, 1995).
31. Virginia P. Dawson, *Engines and Innovation: Lewis Laboratory and American Propulsion Technology* (Washington, DC: NASA SP-4306, 1991).
32. Elizabeth A. Muenger, *Searching the Horizon: A History of Ames Research Center: 1940–1976* (Washington, DC: NASA SP-4304, 1985).
33. Hallion, *On the Frontier*.

focused on a variety of subject areas, such as "theoretical" (often a code word for groups working in computational fluid dynamics), materials, simulation, or safety. The branches were important because they formed the research unit. The branch was where most of the work was performed; it was where new employees were inducted into a particular field of expertise; it was where key technical decisions were made (as opposed to key policy decisions); and it was, for many years, the starting point for climbing the Laboratory hierarchy. The branch system rewarded intellectual rigor and innovative thinking, but not necessarily managerial skills or fealty to NASA Headquarters. Branch chiefs reported to division chiefs, and division chiefs reported to the Center Director (or Engineer-in-Charge, as the position was originally called at Langley). Center Directors in the NACA era exercised significant control over Laboratory budgets. In the NASA era, organizational trends slowly eroded this hierarchy. Branches remained, but increasingly they reported to project offices and the aeronautics directorate operating out of Headquarters. Center Directors, who by the 1970s were starting to come from outside of the branches, became caretakers for various divisions, branches, and facilities that stretched across directorates (e.g., aeronautics *and* space science). The informal, insular, and technocratic traditions of the old Laboratories slowly gave way to remote oversight and increased administrative checks and reporting.

The role of the space program is impossible to ignore in this narrative, for it had a strong impact on the content and organization of aeronautical research. Even if aeronautics is the "other NASA" (and it is probable that other subsets of NASA would also lay claim to being distinct in form and function), it is still part of NASA. As such, it has a shared history as well as a shared leadership and administrative structure with the larger organization. It would be an oversimplification to say that as NASA's space program goes, so goes aeronautics. But in some key respects, the statement holds, and not just because of porous boundaries between the two. First, aeronautics has been subject to long-term administrative trends that began primarily on the big-project, space side of the house, especially the centralization and bureaucratization of research administration. At the risk of overstatement, aeronautics has become a collection of relatively small research programs run with big-program management. Second, aeronautics' budget does not exist within a safe harbor but is subject to Agency-wide pressure to carry out NASA's biggest and most public programs, namely, the human space program. These pressures became particularly acute after the end of the Cold War, which had served as a core justification for military aeronautics research since the late 1940s. Third, for good and ill, aeronautics has ridden the coattails of the space program, sharing in the prestige of the Agency's successes and redoubling its efforts in the face of tragedy and criticism about the Agency's efficacy and goals.

The relationship between aeronautics and the space program is not a question of whether aeronautics would have been better off as a separate entity (a thorny hypothetical), but the ways in which aeronautics research has adjusted to a management structure focused on a very different task. Interestingly, NASA's founders drew a line between the research side of the Agency (which included both space and aeronautics) and the mission-oriented space side. The research side was largely the old NACA community, reconstituted within NASA. For quite a while, this research community continued to do its work as it had under the NACA. With time, this managerial and budgetary wall between research and missions fell, leaving research, including aeronautics research, as just another claimant to NASA's budget. Furthermore, the spread of space-oriented administrative procedures evolved to treat research as an engineering management exercise with predefined goals, metered pacing, and centrally approved budgets. Though these procedures were entirely appropriate for expansive projects that sought to build and deploy space hardware, they were ill fitting to an endeavor that had open-ended possibilities, unknown schedules of discovery and innovation, and hunches that fell outside of budgeted research. In short, big-project management was not the same endeavor as supporting an advanced research community.

Living with the space program had salutary, if unintended, effects for aeronautics research. Notably, making aeronautics second fiddle forced Centers and branches to think about their competitive advantage and to consider alternate technologies and approaches. It is quite possible that the technical narrative would be very different had aeronautics been restricted to the activities of a single Center. One might see in this an evolutionary explanation for the creation and population of new research programs; one might also view this as mere bureaucratic opportunism. Regardless of motivation, the outcome has been fruitful competition and a differentiation of methodologies.

Conversely, this competition for resources led aeronautics to act like the space side of the house, refashioning itself in ways that were not necessarily reflective of either the scale or nature of research in the Labs. In the 1960s, the various aeronautics programs were specifically prevented from overseeing their own SST vehicle program, but come the 1970s, they began inching toward larger policy-directed programs that aggregated numerous small-scale programs. The apotheosis of this trend could be found in the AST and HSR programs of the 1990s. Large, application-oriented programs seemed to have a natural attraction as viable competitors for funding within the Agency and political support on the Hill.[34] The problems for big-name projects were, first,

34. In a 1991 analysis of federal R&D, the CBO noted the attractions of large projects: "Executive agencies may prefer large projects to small projects, because the former provide budgetary

that in their size, they were unrepresentative of the kind of work done at the Centers, and, second, that the Agency simply did not have the resources to give aeronautics a sustained, space-size budget. It remains a continuing challenge for aeronautics to carve from the space agency an organizational apparatus that meets the needs of this "other NASA."

Another inescapable facet of this history is the role played by experimental apparatus and methodology. For the first century of aviation, the wind tunnel has been the archetypal laboratory instrument for aeronautics research. Early on, the NACA developed a close association with wind tunnels, being both innovators in tunnel design and strong proponents of their use. In the 1920s, for example, the NACA established its reputation building and using a pressurized wind tunnel called the Variable Density Wind Tunnel. In the late 1940s, the NACA developed the first useful transonic wind tunnels (i.e., slotted wind tunnels). Wind tunnels were not the NACA's only research methodology; other important approaches included structures, avionics, and flight testing. Yet none matched the importance and utility of wind tunnels. They were logical and flexible instruments, useful for theoretical explorations as well as highly applied studies. More crucially, tunnels allowed researchers to shift back and forth from mathematical models to flight, in the process increasing the reliability of models while also serving to predict aircraft performance. Wind tunnels were, and remain, a critical link between the theory and physical phenomena of flight.

There were also nontechnical reasons for the prominence of wind tunnels. Tunnels fit the American political context; they were general research tools that, for the most part, were too expensive for the private sector. In their high cost to construct and employ, tunnels helped rationalize a continued national investment. Only with World War II did American manufacturers have sufficient resources to begin building their own wind tunnel laboratories, though industry would never attain the breadth of capabilities offered by the NACA and NASA.[35] In short, wind tunnels were easily understood as a public good rather than a

support over a longer period. Large R&D projects offer an executive agency an opportunity to broaden its Congressional support, but along with this support goes a political commitment to keep funding large projects even if cost overruns or shortfalls in agency funding force cutbacks in other R&D spending. Large projects may be favored because of the economic benefits they bring to local communities. Once they are initiated, the momentum of large R&D projects gathers strength from the beneficiaries of project spending in both the public and private sectors." U.S. Congress, Congressional Budget Office, "Large Non-Defense R&D Projects in the Budget, 1980 to 1996," July 1991, p. 42.

35. Ferguson, "Technology and Cooperation."

corporate subsidy, even as they performed research for their corporate patrons. Wind tunnels did not appear to overstep the line between the two.[36]

As noted above, wind tunnels were often tied to the organizational structure of the laboratories, with groups of researchers focused on specific speed ranges and, likewise, with specific tunnels. Taken as a whole, the wind tunnel was not merely an instrument, but a bundled sociotechnical package with strong institutional momentum. It was a research community, a continuous research tradition with built-in mechanisms for the transfer of explicit and implicit information from one generation to the next and between government and private researchers. It encompassed a system for professional advancement. It formed the basis for a long-running funding mechanism that dated to the Progressive Era, and, of course, it was an eminently useful laboratory tool.

An earlier history of the NACA argued (and not without contention from fellow historians) that this momentum behind wind tunnels, behind a particular methodology, overly shaped the direction and administration of research.[37] With so much momentum, it is easy to see that determining which questions to study can become a question of what a tunnel can do and, following from that, how to acquire the next generation of wind tunnel. It comes as no surprise that one of the major battles between the NACA and the U.S. Air Force in the late 1940s and early 1950s was over future wind tunnels (the outcome of which was the Unitary Plan Wind Tunnel).[38] Similarly, the emphasis on wind tunnels and aerodynamics meant less emphasis on other lines of inquiry. Not

---

36. This arrangement with private industry created an important avenue for the flow of technical information to and from the laboratories.
37. Historian Alex Roland, in his history of the NACA, explicitly noted the distorting effects of the Agency's attachment to wind tunnels, writing: "…research equipment shaped the NACA's program fully as much as did its organization and personnel. The NACA achieved early success and acclaim by developing revolutionary wind tunnels for aerodynamical research. Thereafter the tunnels took on a life of their own, influencing the pace and direction of NACA research; concentrating the Committee's attention on aerodynamics when fields like propulsion, structures, and helicopters had equal merit; and becoming in time a sort of end in themselves." Roland, *Model Research*, vol. 1, pp. xiv–xv. It is important to note that Roland's *Model Research* met with disagreement, especially in regard to whether the NACA was alternatively too focused on practical matters or too focused on theory, or whether the Agency was blinkered (and thus missed taking up turbine engines at an earlier point, for example). See William F. Trimble, review of *Model Research: The National Advisory Committee for Aeronautics, 1915–1958*, by Alex Roland, *Isis* 79 (1988): 175–176; and Wm. David Compton, review of *Model Research: The National Advisory Committee for Aeronautics, 1915–1958*, *Technology and Culture* 27 (1986): 653–658.
38. Roland, *Model Research*, vol. 1, pp. 211–221.

until the NACA created the Lewis Engine Laboratory in 1942, for example, did propulsion research begin to approach the priority accorded to traditional tunnel work (and this, perhaps, too late to answer postwar critics who decried the Agency for not having matched the Germans and British in turbine research and rocketry). Likewise, in the 1950s, aircraft manufacturer Douglas argued that the NACA should do for structures and materials what the Agency had done for propulsion: create a separate materials laboratory that was out of the shadow of the aerodynamics core.[39]

For the NASA era, this history digresses from earlier arguments about wind tunnels. While tunnels and aerodynamics remained as crucial pieces of experimental hardware and emphasis, the NASA era is characterized by the growth of alternative methodologies that successfully competed with the tunnels for space and funding at the Labs, even as these methods also complemented traditional experimental means. There are a number of factors behind this evolution. Institutionally, with the switch to NASA and the emphasis on space, wind tunnels were no longer the anchor that they had been. For space, wind tunnels were merely a means to an end. Thus, as with aeronautics generally, wind tunnels dropped a few notches in importance.[40] More fundamentally, the creation of multiple NACA Laboratories, largely an artifact of World War II, sowed the seeds for competing communities of researchers that sought to differentiate themselves through methodology. The prime example of this was Dryden, which showed how a research community could thrive while focusing on flight testing, especially through the use of remotely controlled scale vehicles. Computational fluid dynamics, likewise, became a powerful competitor and complement to wind tunnels. CFD reflected the influence of external (or macro) technological change while it also exemplified the willingness of the Laboratories to seek out new methodologies as a means of differentiation. Finally, we may also point to broader changes in NASA's aeronautics mission: that the Agency and the Laboratories actively sought to attract support by reinventing themselves. Here, the best example is NASA's growing involvement in airspace research. The end result is that at the end of five decades of NASA research, even as wind tunnels remain critical to advanced research, the

39. Edwin Hartman to Director, NACA, 18 July 1951, "Subject: Visits to the Santa Monica Plant of the Douglas Aircraft Company, June 19 and July 2, 1951," Edwin Hartman Memorandums, Langley Research Center Historical Archive. On postwar criticism, see Roland, *Model Research*, vol. 1, p. 204.

40. Interestingly, the space side of NASA has its own need for tunnels (arc jet tunnels and shock tunnels that recreate the conditions of atmospheric reentry), though these represent a very specialized subset of all the Agency's tunnels.

Laboratories have embraced a range of methodologies and have done so in an entrepreneurial fashion.

Organization of the Book

The book is about aeronautics during the NASA era, but to gain the best understanding, we must begin with a discussion of the NACA era. NASA's aeronautics research did not spring forth anew in 1958 but was a continuation of NACA work under a new name and mission. The next chapter examines some of the important long-term characteristics of NACA research, paying close attention to post–World War II developments. This late-NACA period forms a discrete historical unit in terms of technical and political trends. While the NACA's work in the 1920s and 1930s is no less historically important, it was from this post-WWII environment of the Cold War that NASA was born.[41] World War II itself represented a deep discontinuity in technical trends, leading to a postwar emphasis on high-speed, high-altitude research employing jet and rocket engines. The war also, as noted, led to a system of competitive laboratories, something that would have a long-term impact on research strategy.

The NASA chapters (3 to 7) are generally chronological by decade, but in numerous cases, technical programs are discussed out of sequence in order to maintain some of the minor narrative arcs. For example, the Small Aircraft Transportation System (SATS) program ran from 2000 to 2005, yet I grouped it with the Advanced General Aviation Transport Experiments (AGATE), which I cover in the chapter on the 1990s. SATS was a successor program to AGATE, both of them growing from the same concern over the decline of the general aviation market and both of them employing strong collaborative links to private industry. It did not make sense to break this story in two for the sake of tidy chapter divisions.

Chapter 3 begins with the creation of the space agency. Whereas most histories of NASA have focused squarely on space-related R&D, this chapter examines the way in which aeronautics adjusted to the new mandate. For the most part, and in keeping with how the Agency initially was constructed, aeronautics functioned along familiar lines. NASA's management was, after all, preoccupied with larger matters and did not have the resources or time to remake or micromanage

41. The pre-WWII NACA history is ably covered in a number of works, including Hansen, *Engineer in Charge*; and Roland, *Model Research*. Many of the key NACA-era scientific and engineering developments can be found in James R. Hansen, ed., *The Wind and Beyond: A Documentary Journey into the History of Aerodynamics in America*, vol. 1, *The Ascent of the Airplane* (Washington, DC: NASA SP-2003-4409, 2003), and vol. 2, *Reinventing the Airplane* (Washington, DC: NASA SP-2007-4409, 2007).

the aeronautics mission. Still, aeronautics researchers responded to the space race, some jumping entirely into the new enterprise, others crafting research projects that assisted in some fashion. So, for example, there was the X-15 rocket airplane, a continuation of a pre-NASA program, and there was the new lifting-body program, which sought to create flyable reentry vehicles. As noted, NASA conducted research on a supersonic commercial transport, even as it demurred on running a full SST development program. The chapter also covers revolutionary, if prosaic, developments in subsonic wing design.

Chapter 4 investigates the decade of the 1970s, focusing especially on the changes wrought by the drawdown of the Apollo program and external political forces. By this point, NASA's leaders had the time and incentive to focus more attention on aeronautics, and with growing pressure on NASA to make itself relevant to increasing energy and environmental problems, aeronautics offered a distinct opportunity. One result was a large, multifaceted energy-efficiency program for aircraft that drew together a number of existing lines of research. The chapter also explores changes brought about by advances in digital computing, including the rise of computational fluid dynamics and digital fly-by-wire aircraft. The chapter covers work on vertical and short takeoff aircraft, including not only the well-known tilt-rotor experiments, but also work on aircraft meant to operate from short runways, something that was pitched as a potential solution to airspace congestion. Finally, the chapter discusses changes in Center focus, notably the division of labor between Langley and Ames.

Chapter 5 takes the book into the 1980s and the Reagan presidency. It was an era that saw increasing hostility to federally funded research, something that would become a long-term feature of conservative critiques of the government. At the same time, the era saw a renewed emphasis on the Cold War and increased military spending and, ultimately, aeronautical R&D. Whatever the ideological objections to NASA, it was not suffering from a lack of projects. The military push coincided with a rethinking of air combat strategy that focused on stealth and maneuverability and took advantage of earlier developments in digital flight control to make aircraft do things that were hitherto impossible. Among the most remarkable of a string of new X-planes was the X-29, a small jet aircraft with forward-swept wings that relied on computers to stay in the air and, just as important, also depended on an unprecedented understanding of aeroelastic behavior. Another military project that steered work toward NASA was the NASP. Pitched as a commercial technology, the program actually represented an expansion of earlier experimental work on high-speed, high-altitude vehicles for the military. Even if NASP was never close to being realized, it supported fundamental work on materials and propulsion.

The 1980s were not all about Cold War–inspired military experiments. The Advanced Turboprop Project, an heir of the energy-efficiency programs of the 1970s, took wing, as did the AD-1, an oblique-wing aircraft that pivoted its wing about its center. The laboratories also took major steps toward supporting computational flight dynamics and computers as major research methodologies. Ames was the most daring in this regard, creating a Numerical Aerodynamic Simulation (NAS) laboratory that initially sought to build the most advanced computing hardware. At the same time, NASA continued to do the kind of troubleshooting for which it was long known, in this case in the areas of wind shear and icing.

Chapter 6 examines the 1990s, a period in which aeronautics lost one of its main pillars of support, the Cold War. Without a defense imperative, federally funded research became a ripe target for critics, just as funding for military prototypes and X-planes slowed. Commercially, the U.S. aircraft industry was under threat and losing market share to the European airframe maker, Airbus, but this was no replacement for the strategic threat posed by the former Soviet Union. Overlaid on these developments were large-scale changes in how NASA operated and formulated aeronautics policy. NASA Administrator Daniel Goldin sought to make the Agency more nimble and efficient by applying lessons from private industry. Congress introduced complementary measures to make agencies more accountable. The net result was greater and more centralized administrative control over research direction and funding. The Agency repackaged the bulk of its aeronautics research into two programs, High Speed Research and the Advanced Subsonic Technology Program, and was able to reach a funding level of nearly a billion dollars. By the end of the decade, however, the two programs came crashing down amidst Agency-wide funding pressures. In addition to HSR and AST, the chapter covers the UH-60 Airloads program and the expansion of NASA's general aircraft research.

Chapter 7 covers a shorter period, from the turn of the century to the early years of NASA Administrator Michael Griffin. This was a tumultuous period for aeronautics, which, since the late 1990s, jumped from mission statement to mission statement vainly in search of a successful policy formula. While the situation eventually stabilized under Griffin, funding for aeronautics remained drastically lower than it had been at its peak in the 1990s. The chapter discusses three project areas: blended wing body designs, which attempt to provide a revolutionary new configuration for subsonic aircraft; integrated scramjet research that, after decades of study at NASA, lofted a test engine over the Pacific; and air traffic control research, which represented a substantial increase in NASA's purview as well as the culmination of grassroots research begun decades earlier.

Chapter 2:

NACA Research, 1945–58

The history of NASA's aeronautics research correctly begins with the National Advisory Committee for Aeronautics. The NACA formed an organizational platform for the creation of NASA in 1958, while the NACA's focus on aeronautics research formed a subset of the new NASA. As much as NASA represented an organizational revolution and a redirection of R&D effort, the aeronautics core persisted as a fairly continuous strand of history. Though Sputnik proved to be a turning point for the Agency and spaceflight, the end of World War II was similarly critical for aeronautics, for the war established a set of political, organizational, and research challenges that persisted through the second half of the 20th century. An outline of the NACA between 1945 and 1958 establishes a baseline for understanding the impact of the space program and brings into relief long-term changes within postwar aeronautics.

Among the challenges of the NACA after World War II was maintaining its position within a crowded market for federally funded research. Though its budget was larger than ever, it had to bargain for resources and agendas shared with the military and industry. In this, the definition of NACA research existed as a political compromise, fitting awkwardly amid the desire to produce long-term public goods, the search for military superiority, and support for commerce. While the NACA suffered considerably for its perceived lapses in World War II, it retained political backing and continued to receive relatively unhindered support for four communities of aeronautical researchers at Langley, Ames, Lewis, and the High Speed Flight Station (HSFS, later renamed Dryden).

With steady leadership and an organizational structure characterized by engineering committees and technical peer review, the NACA offered a safe harbor for the exploration of the multitude of problems facing postwar aviation. Making up for wartime shortcomings vis-à-vis German laboratories, the NACA focused predominantly on high-speed research and jet turbines. It established very close ties to the military, especially the Air Force, through the joint development of experimental vehicles. In engine research, it formed a hands-on relationship with the engine manufacturers. Spaceflight was a peripheral activity before the launching of Sputnik, but by January 1958, the NACA leadership chose to recast the organization's mission.

The Postwar NACA

The full impact of World War II and the Cold War came late to the NACA. While the NACA grew during the war and was extremely busy testing military designs, the magnitude of its growth was never enough to prompt widespread organizational change. The wartime NACA was indeed much larger, going from 500 employees in 1939 to 6,077 in 1945, but it retained the same committee structure that had characterized the organization since World War I.[1] Similarly, though its research shifted away from knowledge production and toward the refinement of military vehicles during the war, it conducted its work in much the same fashion as it had before mobilization.[2] To borrow an engineering phrase, the NACA was scalable. Wind tunnels and personnel could be multiplied without substantially changing the organization's structure or routines. This was not true of either the military or industry. By war's end, the production of new knowledge and leading-edge hardware fundamentally had changed the aircraft industry and military R&D. The relevance for the NACA was twofold.

First, the NACA had not been pressed to acquire the same kinds of development capabilities as the military or industry. Aircraft companies, for example, were not simply larger than they were before the war; they were now able to rapidly develop new technologies across large geographic and industrial spaces using knowledge that was both internally and externally generated. Crucially, aircraft companies were reaching out to technologies that were well beyond the boundaries defined by aeronautical science and engineering. Not until Sputnik would the NACA (and then NASA) be faced with similar developmental, cross-disciplinary, project-oriented challenges. Indeed, this would prove to be a major focus of NASA Headquarters in the early years of the Agency; by the end of the Apollo program, NASA would be known as one of the world's leading innovators in the kind of systems management and engineering that

1. Roland, *Model Research*, p. 489.
2. Illustrating the level of integration between the NACA and the military was the fact that Langley received an early production copy of each new aircraft in order to test for control and performance. See W. Hewitt Phillips, *Journey in Aeronautical Research: A Career at NASA Langley Research Center*, Monographs in Aerospace History, no. 12 (Washington, DC: NASA, 1998), pp. 62–63.

underpinned large-scale, high-technology development.[3] For the 1940s and early to mid-1950s, however, the NACA was still operating much as it had before World War II.

Second, the NACA after World War II was no longer the curious and atypical federal research program of the prewar years.[4] The end of World War II and the rise of the Cold War made scientific discovery and technological innovation federal mandates. The NACA was the big fish in a small pond before World War II, sharing its space mainly with universities and the odd industry-built wind tunnel. After the war, the NACA occupied a much smaller niche, not only within the universe of different disciplines such as atomic energy and electrical engineering, but also within aeronautics. Aircraft companies and their associated suppliers in the engine and electronics fields had become sophisticated R&D facilities in their own right. The NACA still had the most advanced wind tunnels and some of the brightest aeronautical researchers, but it was now one of numerous knowledge producers contributing to the advancement of aviation, military and commercial.[5]

---

3. Stephen B. Johnson, *The Secret of Apollo: Systems Management in American and European Space Programs* (Baltimore: Johns Hopkins University Press, 2002).
4. Not that federally funded research was rare, for it long had been found in such places as the Bureau of Standards, the Department of Agriculture, the Coast and Geodetic Survey, and the military. What distinguished the NACA was its focus on such a narrow and highly applied technological field. Such programs would become more commonplace after World War II (e.g., support for electronic computing). See Dupree, *Science in the Federal Government*; and David C. Mowery and Nathan Rosenberg, *Technology and the Pursuit of Economic Growth* (Cambridge, U.K.: Cambridge University Press, 1989). On the growth of federally funded science after World War II, see Vannevar Bush, *Science, the Endless Frontier* (Washington, DC: U.S. Government Printing Office, 1945); Daniel J. Kevles, *The Physicists: The History of a Scientific Community in Modern America* (Cambridge, MA: Harvard University Press, 1987); Peter Galison and Bruce Hevly, eds., *Big Science: The Growth of Large-Scale Research* (Stanford, CA: Stanford University Press, 1992); and Stuart Leslie, *The Cold War and American Science: The Military-Industrial-Academic Complex at MIT and Stanford* (New York: Columbia University Press, 1994).
5. NACA Chairman Jerome Hunsaker recognized as early as 1944 the importance of this shift in relative research capacity. His "Memorandum on Postwar Research Policy for NACA," 27 July 1944, listed the growth of military and industrial research facilities around the country. Memo reprinted in Roland, *Model Research*, Appendix H, document no. 34, pp. 684–686. Aside from the sizable expansion of industrial R&D facilities during and after World War II, a good example of the sophistication of industrial R&D can be found in Edwin Hartman's reports (Hartman was the NACA west coast liaison). In a 27-page memo written in 1952, Hartman went into great detail to describe Boeing's advanced research into flutter. Boeing had the largest group working on the

The perceived success of strategic bombing and carrier aviation in World War II, as well as the consequent rise in the aircraft industry's power and influence, put the NACA on a new political footing. Even as aviation finally received the validation its promoters had been seeking for over three decades, the NACA was unable to retain its former stature. The air arms of both the Army and Navy, of course, emerged from World War II with sufficient momentum that the country would authorize a separate air force in 1947 and orient naval operations around carrier groups rather than battleships. More importantly, the air arms not only increased their support for in-house aeronautical R&D but also nurtured new sources of scientific and technical advice that made for an increasingly crowded space in the field of aviation expertise. General Henry "Hap" Arnold's efforts are notable in this regard. In 1944, Arnold formed his own Science Advisory Group (SAG) under the leadership of Caltech's Theodore von Kármán; this would become the Science Advisory Board (SAB) in 1946 (a group that functions to this day within the U.S. Air Force).[6] Arnold also instigated changes that gave industry a stronger voice in research leadership; in 1946, he called upon his friend Donald Douglas, president of Douglas Aircraft, to help consult on research matters for the Air Force, an effort that

---

issue at the time (29 people), and other manufacturers were copying Boeing's methods. Edwin Hartman to Director, NACA, 10 March 1952, "Subject: Visit to the Boeing Airplane Company, Seattle, February 18, 19, and 20, 1952," Edwin Hartman Memorandums, Langley Research Center Historical Archive. Vic Peterson, a longtime Ames researcher, noted that while industry did have wind tunnels, they lagged behind the NACA's capabilities, and they were generally much smaller (due to the cost of construction and power consumption). See Victor L. Peterson, interview by Robert G. Ferguson, tape recording, Los Altos, CA, 17 January 2005, copy located at the NASA Historical Reference Collection, NASA History Program Office, NASA Headquarters, Washington, DC (hereafter "HRC").

6. Under von Kármán and with the assistance of seed money from the Daniel Guggenheim Fund for the Promotion of Aeronautics, Caltech's aerodynamics program was the NACA's closest rival in the U.S. before World War II. On the SAG and SAB, see Thomas A. Sturm, *The USAF Scientific Advisory Board: Its First Twenty Years, 1944–1964* (Washington, DC: U.S. Air Force [USAF] Historical Division Liaison Office, 1967). See also Jacob Neufeld, ed., *Reflections on Research and Development in the United States Air Force: An Interview with General Bernard A. Schriever and Generals Samuel C. Phillips, Robert T. Marsh and James H. Doolittle, and Dr. Ivan A. Getting*, conducted by Dr. Richard H. Kohn (Washington, DC: Center for Air Force History, 1993). On von Kármán, see Michael H. Gorn, *The Universal Man: Theodore von Kármán's Life in Aeronautics* (Washington, DC: Smithsonian Institution Press, 1992); and Paul A. Hanle, *Bringing Aerodynamics to America* (Cambridge, MA: MIT Press, 1982).

would eventually become the RAND Corporation.[7] Even within the NACA, these wider political shifts became reflected in the makeup of its committees. When the NACA was founded in 1915, it was done with a wary eye toward the influence of industry, which, in the tradition of the Progressive Era, was assumed to be corruptive. The NACA's governance relied primarily on the advice of civil servants, academics, and scientifically minded members of the Army and Navy. Mobilization for WWII, however, made acceptable the close ties between government and industry (e.g., the rise of "dollar-a-year men," industry leaders who worked on behalf of the government or military). With the NACA under pressure in Washington, DC, Chairman Jerome Hunsaker reached out to industry. The NACA created a high-level advisory group, the Industry Consulting Committee (ICC), composed of the leaders from the largest airframe and engine manufacturers. Among the ICC's first actions was to recommend increased representation on the NACA Main Committee, which it got (increasing from 1 to 3 out of 15).[8]

In this context of both uncertainty and opportunity, the NACA set out to encourage greater support for aviation research and maintain a position of leadership. Unlike the situation after World War I, when the NACA was charged with bringing the United States to the forefront of aeronautical research, the United States after World War II held a clear technological edge over most other nations—if not in research, than at least in resources. At the same time, the war had exposed the degree to which the NACA had, at least to observers, failed to explore promising technologies such as jet propulsion, swept-wing aircraft (in spite of having made their own independent discoveries), and guided missiles.[9]

---

7. Dik Alan Daso, *Hap Arnold and the Evolution of American Air Power* (Washington, DC: Smithsonian Institution Press, 2000); Martin J. Collins, *Cold War Laboratory: RAND, the Air Force and the American State, 1945–1950* (Washington, DC: Smithsonian Institution Press, 2002); Dik Alan Daso, "Operation LUSTY: The US Army Air Forces' Exploitation of the Luftwaffe's Secret Aeronautical Technology, 1944–45," *Aerospace Power Journal* 16, no. 1 (spring 2002): 28–40.
8. Roland, *Model Research*, pp. 201–211, 431–435.
9. Both Roland in *Model Research* and Hansen in *Engineer in Charge* discuss the question of whether the NACA missed the boat. Historiographically, it is important to ask whether jet engines, swept wings, and guided missiles were part of a logical sequence that could be anticipated or whether German and British developments were merely accidental and fortuitous. In any case, for my purposes here, it is enough to note that some participants did believe that the NACA had not kept up with its British and German peers before the war. See Roland, *Model Research*, pp. 204–205. Writing four decades afterward, Langley researchers noted that the NACA had, indeed, been performing swept-wing research and had made similar discoveries, independently, five years after the Germans: see Edward C. Polhamus and Thomas A. Toll,

In typical fashion, the NACA established the Special Committee on Post-War Aeronautical Research Policy and, in cooperation with the Army Air Forces, the Navy Bureau of Aeronautics, the Civil Aeronautics Commission, and the Industry Consulting Committee, drafted a policy statement in March 1946. In sentiments that would be echoed through the Cold War, the committee warned of the danger posed by losing international research supremacy: "The effects of accelerated enemy research and development in preparation for war helped to create an opportunity for aggression which was promptly exploited. This lesson is the most expensive we ever had to learn. We must make certain that we do not forget it."[10] As to the place of the NACA, the committee reiterated what had long been the division of labor among the NACA, the military, and industry: "Fundamental research in the aeronautical sciences is the principal objective of the NACA."[11] This formulation was meant to protect the military's own aeronautical research interests while keeping the federal government from unnecessarily subsidizing (or competing with) private industry. It was a political expedient that allowed for a civilian agency to conduct technological research for the public good, but it was a formulation fraught with pitfalls. If the NACA strayed too close to the domain of private research, manufacturers or members of Congress might object. Yet if the NACA veered too far into areas of dubious practical application, again, manufacturers and members of Congress might object. As a public agency, the NACA was meant to create knowledge that was a public good, unbeholden to any private interest, and to operate at a level above and before the design process of the manufacturers. Yet the NACA, Congress, the military branches, and the manufacturers worried that public distribution of the NACA's reports hindered industry's ability to capture the country's research investment and maintain American technological superiority. In spite of all the talk about the NACA's conducting fundamental research, its studies ran the gamut from pure to applied research, and its patrons in the military and industry had a voracious appetite for all manner of investigations, fundamental or otherwise.[12]

Fundamental aeronautical research was not a mere rhetorical ploy, but a politically defined space for scientific and engineering activities. Certainly, the NACA sought to sit atop the research pyramid (by performing the most

---

"Research Related to Variable Sweep Aircraft Development" (Langley Research Center, Hampton, VA: NASA TM-83121, May 1981).

10. NACA, "National Aeronautics Research Policy," 21 March 1946, folder 15861, NASA HRC.

11. Ibid.

12. For an example of the threat that the NACA posed to industry, see Dawson, *Engines and Innovation,* p. 81.

basic research) and to coordinate the country's research activities by sitting at the nexus of various scientific and engineering communities. Unfortunately, the NACA would have neither a lock on fundamental research nor a commanding position in the postwar environment. Indeed, no single organization would because resources and technological fronts multiplied and shifted. The full impact of the NACA's diminished role is perhaps best understood in its contribution to the formation of air policy in the executive branch. Whereas the NACA was created in 1915 to serve as *the* consultative body for the government on matters aeronautical, President Harry S. Truman sidestepped the NACA in 1947 when he sought advice about postwar aviation. He called upon Thomas K. Finletter, an attorney who had served as Special Assistant to the Secretary of State from 1941 to 1944, to head the President's Air Policy Commission (known widely as the Finletter Commission). No member of the NACA served on the commission, though they did testify, and in the critical field of research and development, the commission's advisor was Grover Loening, a former aircraft manufacturer. The NACA was now one voice among many and, from the government's perspective, one of a number of claimants for aviation research funds.[13]

The Finletter Commission's report, *Survival in the Air Age*, placed aeronautical R&D front and center. It made clear that technological change in aviation was responsible for the nation's new security vulnerabilities as well as a keystone of national defense. Like Paul Nitze's NSC 68 and Vannevar Bush's *Science: the Endless Frontier*, the commission's report was a kind of Cold War manifesto, though it was less an influential document than a reflection of views held by aviation's elite. Presciently, the commission anticipated a policy of nuclear deterrence by arguing for aeronautical capabilities sufficient to overwhelm any aggressor:

> We also must have in being and ready for immediate action, a counteroffensive force built around a fleet of bombers, accompanying planes and long-range missiles which will serve notice on any nation which may think of attacking us that if it does, it will see its factories and cities destroyed and its war machine crushed. The

13. Of the 150 individuals who testified before the commission, 3 were from the NACA (as many as from Douglas Aircraft Co.): Jerome Hunsaker, Hugh Dryden, and John Victory. The commission visited the NACA's Langley and Ames Laboratories, but it also visited 16 manufacturers. President's Air Policy Commission, *Survival in the Air Age* (Washington, DC: U.S. GPO, 1948), pp. 158–166. The Finletter Commission's papers are kept at the Harry S. Truman Presidential Library in Independence, MO.

> strength of the counteroffensive force must be such that it will be able to make an aggressor pay a devastating price for attacking us.[14]

The Commission recommended that "military security must be based on air power."[15] This air power would rely not on simple numerical supremacy (as had often characterized pre–World War II debates about preparedness), but on technological advantages born of research and development.

The Finletter Commission restated the NACA's titular role in the country's R&D pipeline: the NACA would perform fundamental research while the Army, Navy, and manufacturers carried out development tasks. At the same time, however, the Commission indirectly recognized that aeronautics was moving into areas that were well outside the NACA's traditional competence, including electronics, atomic energy, missiles, and jet propulsion. The NACA clearly occupied a precarious niche tied not so much to fundamental research (whatever that was), but to its relatively focused aerodynamic expertise and its collection of test instruments.[16]

Taken as a whole, the NACA began the postwar era with less political power than it had before, with an organization that was markedly less sophisticated than its competitors in the military and industry, and with a research focus that, relatively speaking, had become narrower. Though far from being ineffectual or doomed, the NACA was nonetheless increasingly fighting for its share rather than nobly carrying the torch of aeronautics for the nation as it had done in the 1920s and '30s. Aviation's place in both defense and commerce had been secured. Still, the NACA's support was sound, resting not just on its highly respected tradition of research, but on the quick pace of technological development occurring in the aircraft industry and especially on the omnipresent threat of the Cold War. Regardless of how politics defined the NACA's research boundaries and irrespective of ideological attitudes about federally funded research, the NACA and, later, NASA could expect continued patronage so long as the three pillars of industrial support, national security, and rapid technological change remained intact.

Laboratories and Leadership

Prior to becoming NASA, the NACA operated three laboratories and two flight-test areas. The oldest was the Langley Laboratory near Hampton, Virginia. Formally dedicated in 1920, Langley was the Committee's only facility until

---

14. Ibid., p. 20.

15. Ibid.

16. Ibid., pp. 8, 71–97.

1940, and it remained the patriarch among the Laboratories for quite some time. At its pre-NASA height in 1952, Langley employed 3,557 people and, from 1920 to 1958, had constructed some 30 wind tunnels (of which about half were still operational in the late 1950s), 2 towing tanks, and various specialized laboratories. It occupied over 700 acres of land next to the Back River and shared some facilities, such as runways, with the adjoining Langley Air Force Base.[17]

The second laboratory to open was Ames Aeronautical Laboratory, next to the Navy's Moffett Field in Sunnyvale, California. Ames was, at least in the beginning, a West Coast Langley, the two Labs having similar test equipment and research functions, especially during the war. A contingent of scientists and engineers from Langley moved to California to build and operate Ames. To some extent, Ames was a rearguard action meant to keep the NACA relevant to an industry that had shifted from the two principal East Coast aircraft manufacturing centers, Buffalo and Long Island, New York, to the West Coast.[18] From 1945 to the late 1950s, Ames's budget was usually less than half that of the Langley Lab's, and it employed half as many personnel. Its facilities included 16 wind tunnels constructed from 1940 to 1956.[19]

The third laboratory was the Lewis Engine Lab, opened in 1942 on 200 acres of land outside Cleveland, Ohio, and next to the Cleveland Municipal Airport (now Cleveland Hopkins). It was in regional proximity to the Army Air Force's Power Plants Laboratory in Dayton and relatively close to the nation's aircraft engine industry. The dominant reciprocating engine companies were Wright (Paterson, New Jersey), Pratt & Whitney (East Hartford, Connecticut), and Allison (Indianapolis, Indiana). The major jet engine companies were General Electric in Lynn, Massachusetts, and Cincinnati, Ohio (1949); Westinghouse in Philadelphia, Pennsylvania, and Kansas City, Missouri; and Pratt. The Lewis Lab was a direct descendant of Langley's Aircraft Engine Research Laboratory, begun in 1934. By the late 1950s, the lab was employing over 2,700 persons and its budget nearly rivaled Langley's. Though Lewis had far fewer wind tunnels than Langley, propulsion research was nonetheless very expensive.[20]

---

17. The history of the Langley Lab in the NACA years has been ably described in Hansen's *Engineer in Charge*.
18. The major West Coast manufacturers included Boeing, Consolidated-Vultee, Douglas, Hughes, Lockheed, North American, Northrop, and Ryan. See especially the primary documents reproduced in Roland, *Model Research*, pp. 678–684.
19. Roland, *Model Research*, table C-6, p. 480, and appendix E, pp. 507–528; Elizabeth A. Muenger, *Searching the Horizon: A History of Ames Research Center: 1940–1976* (Washington, DC: NASA SP-4304, 1985), appendix C, p. 233.
20. Dawson, *Engines and Innovation*, appendices B, D, and E.

Figure 2.1. NACA Appropriations, Actual Dollars

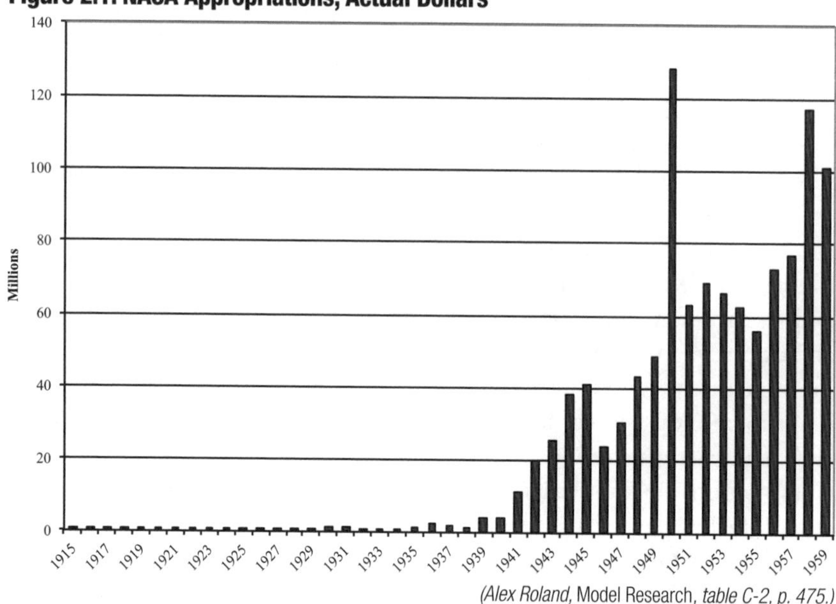

(Alex Roland, Model Research, *table C-2, p. 475.)*

The spike in funding in 1950 was due to additional construction costs for new testing facilities.

The two test areas included Wallops Island (also known as the Pilotless Aircraft Research Facility) and the High Speed Flight Station (HSFS, also known as NACA Muroc and later renamed in honor of Hugh Dryden in 1976). Wallops, on Virginia's Atlantic shoreline, began operations in 1945 and took direction from Langley. The HSFS was located on the Muroc Dry Lake, northeast of Los Angeles, California. Both facilities were, by design, sufficiently remote that they could conduct flight experiments without compromising the safety of people and property. At Wallops, researchers launched rockets over the Atlantic Ocean; at the HSFS, test pilots could take advantage of the clear Southern California skies and the wide expanse of dry lakebed that afforded ample emergency landing opportunities. Additionally, the HSFS could conduct flight experiments in relative privacy. Funding for both Wallops and the HSFS was never large; by 1958, Wallops's budget was $2.3 million and the HSFS's budget was $2.6 million. As test facilities, they were usually completing projects that had begun at one of the laboratories (and thus were accounted for under separate budgets).[21]

All of the Laboratories operated with a minimum of centralized oversight. Not only did they benefit from a geographic distance from Washington, DC,

---

21. Roland, *Model Research*, table C-6, p. 480.

but the Laboratories also operated within an organizational structure that was, by the 1950s, becoming anachronistic. While the structure was hierarchical, it also incorporated layers of advisory committees made up of scientists and engineers. Truly, oversight depended not so much on administrative mechanisms, but on the character and initiative of its leadership. At the heart of the NACA was the Main Committee, initially a group of 12, which served as a board of directors. In the beginning, it was composed of representatives from the Army (2) and the Navy (2), the Secretary of the Smithsonian, the Chief of the Weather Bureau, the Director of the National Bureau of Standards, and five members from "private life." The committee met at least semiannually and served primarily to provide long-range guidance, approve major research programs, and make key decisions about leadership positions. Day-to-day management fell upon the Executive Committee in Washington, DC, and the Engineers-in-Charge at the Laboratories. Personalities figured large in the NACA's history. Two members of the Main Committee, who also chaired the Executive Committee, had an outsize influence on the quality and tone of the NACA: Joseph S. Ames and Jerome C. Hunsaker. Both men were highly regarded, and both were willing to stake their reputations standing up for the NACA.

Ames was a professor of physics at Johns Hopkins University. He served on the Main Committee from 1915 to 1939 (chairman from 1927 to 1939) and as chairman of the Executive Committee from 1920 to 1937. Ames became president of Johns Hopkins in 1929 but continued to commute once a week from Baltimore to Washington, DC.[22] Ames was able to play the roles of both outsider and insider. Hunsaker, even before becoming chairman, was an instrumental figure in American aviation. A naval officer who was one of MIT's first graduates in aeronautical engineering, he was a perennial advisor and sometime critic of the NACA under Ames. Hunsaker was head of the Aeronautical Engineering Department at MIT from 1936 to 1951, and, like Ames, he split his time between his university obligations and his duties at the NACA. He served as chairman of both the NACA and the Executive Committee from 1941 to 1956, replacing his MIT colleague, Vannevar Bush, who led the NACA briefly after Ames's departure.[23]

22. Roland, *Model Research*, p. 101; Henry Crew, "Biographical Memoir of Joseph Sweetman Ames, 1864–1943," in the National Academy of Sciences *Biographical Memoirs*, vol. 23 (Washington, DC: National Academy of Sciences, 1944).
23. For information on Hunsaker, see Roland, *Model Research*; and William F. Trimble, *Jerome C. Hunsaker and the Rise of American Aeronautics* (Washington, DC: Smithsonian Institution Press, 2002).

The other three principal individuals at Headquarters were George W. Lewis, Hugh L. Dryden, and John F. Victory.[24] Lewis came to the NACA in 1919. He had graduated from Cornell with a master's in mechanical engineering, worked as engineer-in-charge at a Philadelphia research foundation, and served as a member on an NACA technical subcommittee. Lewis began as executive officer and, within five years, became the director of aeronautical research, a position he held until 1947. George Lewis was not simply an influential leader; he was for much of the NACA's history synonymous with the organization. He carried it through its formative years; he oversaw the organization during some of its most productive research; and he led the NACA through the difficult war years. His replacement, Hugh Dryden, came to the NACA after a prolific career at the National Bureau of Standards, where he headed the Aerodynamics section. A physicist educated under Joseph Ames at Johns Hopkins, Dryden was exceptionally smart. Lewis handpicked Dryden, and the latter served until the NACA became NASA. Dryden would be very influential in reformulating the NACA's mission for the Space Age.

John F. Victory is worth mentioning in this abbreviated list because he served as the guiding administrative officer for much of the NACA's history. Victory was a stickler for proper administrative procedure as well as an undaunted lobbyist for the organization. He nearly single-handedly kept Headquarters and the Laboratories up to date with their paperwork, budgets, and political obligations. The NACA was a highly regarded government agency not only because of the stature of its research, but because Victory could be trusted to return unspent appropriations to the Treasury. Such conscientiousness and professionalism attenuated but hardly eliminated outside demands for oversight and reform of the NACA's nebulous technocracy.

As noted above and in NACA histories, the Laboratories benefited from little oversight and significant administrative distance from Washington, DC. This is not to say that the Laboratories operated with autonomy. What freedom they had they enjoyed because the personalities in Washington granted it to them. This is an important point, especially when considering developments in the NASA period when the Laboratories began to lose that precious autonomy. The potential for centralization had always existed; autonomy had never been a structural feature of the organization. Thus, changes at Headquarters could and would ultimately bring fundamental changes to the conduct of Laboratory research in the NASA years.

---

24. Roland's *Model Research* provides the best understanding of these figures and their roles at the NACA.

In the NACA years, however, there would be remarkably few changes at the Laboratories. With promotions typically occurring from within the organization (and within the Laboratory), as well as long leadership tenures, the Labs enjoyed years of consistent administration. At Langley, electrical engineer Henry J. E. Reid became Engineer-in-Charge in 1926, and he ran the Laboratory until 1958 (his position later renamed simply Director). At Ames, Smith J. DeFrance, who was Langley's Assistant Chief of Aerodynamics, became the Lab's first Engineer-in-Charge and, likewise, remained at the helm until 1965. The Lewis Laboratory had an administrative head, Edward R. Sharp, who managed the facility from 1942 until 1958. Abe Silverstein, also an aerodynamicist from Langley, effectively oversaw research at Lewis, first serving as the Chief of Research from 1950 to 1952 and then as Associate Director until 1958. He became Lewis's Director in 1961.

Research decisions were largely a matter for the Laboratories. While Headquarters had authority over the funding of research programs, it gave broad authority to the Labs to choose their own direction and, once granted appropriations, disburse their own funds. Writing of the pre–World War II era, Roland notes, "The headquarters was needed to secure funds, mend political fences, prevent duplication, and keep the Langley program in line with the needs of the NACA's customers, especially the military services and the aircraft industry."[25] The Engineers-in-Charge cherished and protected their authority. At Langley, Henry Reid required that all outgoing communication go through his office.[26] With the growth of the NACA during World War II and afterward, Headquarters was simply incapable (and unwilling) to control or manage research at the Labs. The level of decentralized decision-making was likely at its apex in the 1950s.

The NACA supplemented its formal hierarchy with a system of layered committees that advised on technical matters. Research oversight, such as the committees offered, was, in theory, distinct from political and administrative support. The principal committees that answered to the Main Committee focused on the areas of aerodynamics, power plants, and aircraft construction. There were additional committees of varying tenure. Perhaps the most notable was the Industry Consulting Committee that operated from 1945 to 1958; its emergence reflected the larger, postwar political shift noted above. At the next committee layer were the subcommittees, whose specialties ranged from radiator design to high-speed aerodynamics. Committees at all levels were meant to bring together the top experts in each particular realm. In 1948, just

25. Roland, *Model Research*, p. 101.

26. Hansen, *Engineer in Charge*, p. 32.

to pick one year, there were 78 persons serving on committees and 271 serving on subcommittees. Respectively, NACA staff represented 15 and 14 percent, military personnel represented 19 and 25 percent, and members from private industry represented 44 and 39 percent. The ratio of the three groups would change only slightly 10 years later, with industry increasing its membership to 50 and 44 percent.[27]

Whether establishing a standard, defining best practice, or deciding on the best course of research, the NACA committees and subcommittees operated in much the same fashion. The NACA sought to gather the brightest people from a particular field, and these individuals were to approach their respective problems as a technocracy united by a common goal. As with industry, the NACA's committee structure was an organizational compromise, one that allowed it to balance the competing interests of the military branches, government agencies, and industry. By giving everyone a voice at the table, it was less threatening than a closed, hierarchically arranged research organization that pursued topics as it saw fit. With committees, the NACA could avoid stepping on toes and endangering its political support. Hand in hand with its "fundamental research" policy, the NACA's structure and objectives were tailored for survival in its political niche.

While the committee structure was straightforward, its relationship to the NACA's decision-making process was more ambiguous (or, alternatively, flexible). The NACA leadership did not require committee approval for new research programs. It was largely up to the directors and the Executive Committee to ask for advice when the organization truly required assistance or when committee decisions might provide some political advantage or cover. Thus, a progressive technocracy was married to a modest, personality-driven hierarchy. The loose coupling between the committees and the leadership also characterized the committees' role in peer review. The formal products of the NACA's Laboratories were reports. The oversight procedures instituted by George Lewis began with a review of the paper within the Laboratory, followed by reviews from the appropriate technical committee. It was largely the responsibility of the author or authors to make the requested changes in the research or the paper. Under Dryden, peer-review procedures were left up to the individual Laboratories.[28]

There are many good things to be said of the NACA's leadership and structure. The consistency of operations meant that researchers had predictable

27. See Roland, *Model Research*, appendix B, pp. 423–466, for Committee information.

28. See Hansen, *Engineer in Charge*, pp. 35–36; Roland, *Model Research*, p. 244; and Phillips, *Journey in Aeronautical Research*, pp. 160–162.

levels of support and could invest more of their time in research. Oversight, from the researcher's perspective, tended to serve as a positive influence and was not tightly coupled to future funding and administrative procedures. Research output *did* influence one's professional standing within the Laboratory; leadership positions, small and large, tended to go to insiders who had earned their peers' respect. The Laboratories were thus highly competitive communities, but not because of any administrative checks on productivity. Research output was not closely metered, nor were budgets tightly administered. The Labs thus provided great freedom within an intellectually challenging environment.[29]

While in hindsight NACA researchers appeared to be largely unburdened by administrative chains, there was still a trend, even in the NACA days, toward greater oversight. By the mid-1950s, for example, the NACA was already feeling the effects of more widespread managerial innovations and had implemented a Management Control Information System (MCIS).[30] Where George Lewis had eschewed the creation and publication of organizational charts, the 1950s saw the slow encroachment of formalized and explicit administrative procedures.[31] Congress, for its part, was encroaching on the NACA's treasured isolation by beginning to demand annual budgets.[32] Perhaps most importantly, as historian Alex Roland pointed out, the NACA's committee-based structure was out of favor in the new era of federally funded science. Even as the NACA persisted in the 1940s and '50s, it was going against the grain of large research programs managed according to contemporary organizational models.[33]

Tunnels, X-Planes, and Space

Adding to the embarrassment of meeting German jet fighter aircraft in the sky and learning of British achievements in jet engines during World War II,

29. Langley engineer H. Hewitt Phillips writes of his NACA years, "During the entire time that I worked in the Flight Research Division, from 1940 to almost 1960, I never had to worry or even think about the money required to do a job.... To obtain approval for most projects, it was necessary only to have a talk or write a memorandum for the Section Head.... Evidently, the flight research work conducted at Langley was quite expensive, but I did not know the process for obtaining the necessary funds.... Keeping track of the progress or completion of a project was equally informal." Phillips, *Journey in Aeronautical Research*, p. 162.
30. For reference to the MCIS, see E. H. Chamberlin, EO, NACA General Notice, "List of NACA Management Manual Issuances," 5 April 1955, folder 15861, NASA HRC.
31. See, for example, Howard N. Braithwaite (Assistant Classification and Organization Officer), "Functional Statement Chart," July 1954, folder 15861, NASA HRC.
32. Roland, *Model Research*, pp. 221–222.
33. Ibid., pp. 202, 268–269.

the NACA and the U.S. military were taken aback by Germany's aerodynamic research infrastructure.[34] To the teams of American scientists, engineers, and military personnel who gained early access to these facilities in the spring of 1945, the sophistication and advanced state of German research were plainly evident despite the ruinous effects of war. In addition, Germany had plans for a next stage of aeronautical research involving very high-speed, high-power wind tunnels that anticipated the operating conditions of jet- and rocket-powered vehicles. Both the NACA and the Army Air Forces returned home with new ideas about where the U.S. needed to invest its research dollars.[35]

Crucially, the U.S. Army Air Forces (USAAF) drew up plans in 1945 for a competing research organization, which would become the Air Engineering Development Center (AEDC, later renamed the Arnold Engineering Development Center) in Tullahoma, Tennessee. The programmatic vision was laid out in Theodore von Kármán's *Where We Stand*, delivered to Hap Arnold in August of 1945, and detailed further in the multivolume, multiauthor study *Towards New Horizons*. Hugh Dryden, incidentally, was both a contributor and an editor for these works, though this was before he entered the employ of the NACA. Von Kármán was explicit about the need for the service to have its own integrated R&D program:

> Leadership in the development of these new weapons of the future can be assured only by uniting experts in aerodynamics, structural design, electronics, servomechanisms, gyros, control devices, propulsion, and warhead under one leadership, and providing them with facilities for laboratory and model shop production in their specialties and with facilities for field tests. Such a center must be adequately supported by the highest ranking military and civilian leaders and must be adequately financed, including the support

34. On the American perceptions of German jet engine research, see Virginia Dawson, "The American Turbojet Industry and British Competition: The Mediation Role of Government Research," in *From Airships to Airbus: The History of Civil Commercial Aviation*, vol. 1, *Infrastructure and Environment*, ed. William M. Leary (Washington, DC: Smithsonian Institution Press, 1995), p. 129.
35. Theodore von Kármán described eight different supersonic wind tunnels located in Kochel, Göttingen, Braunschweig, and Aachen, the fastest being a 16-by-16-inch tunnel that operated past Mach 4. In development were five new tunnels at Kochel, Göttingen, and München, the fastest of which was to reach Mach 7 and have a test section of 40 by 40 inches. Theodore von Kármán, *Where We Stand: A Report for the AAF Scientific Advisory Group* (Dayton, OH: USAAF Air Materiel Command, 1946), pp. 3–4.

> of related work on special aspects of various problems at other laboratories and the support of special industrial developments.[36]

Von Kármán supplied a willing Hap Arnold with a scientific rationale for a new air force based on R&D. The USAAF's embrace of science and comprehensive weapons development threatened to marginalize the NACA's role. To the NACA's longstanding charge of performing fundamental research, the USAAF's decision added a new, unofficial goal in the postwar era: pursuing research and facilities that strategically defended the NACA. The battle for the next generation of high-speed wind tunnels, as inspired by the German program, played out in what would become the Unitary Plan Wind Tunnel.

The NACA's competing vision was its Supersonic Research Center. Jerome Hunsaker, anticipating that a postwar Congress would not fund duplicate programs in both the NACA and the military, called for coordination among the different parties. Under the leadership of Arthur E. Raymond, the Chief Engineer at Douglas Aircraft (a company with long ties to Hap Arnold), the NACA's special panel on supersonic laboratory requirements defined a "unitary" plan that included both the NACA's National Supersonic Research Center and the USAAF's AEDC. Raymond's panel merely unified everyone's requests; it did little to divide responsibilities. After three years of political arm wrestling, the Raymond plan emerged as the Unitary Wind Tunnel Plan Act of 1949. The NACA was denied its Supersonic Research Center and, instead, allowed to build three high-speed wind tunnels that were to serve, officially, the needs of industry. The newly minted Air Force, however, received initial funding for its AEDC. As historian Alex Roland noted, the NACA emerged from this battle third behind industry and the military.[37]

It would be the mid-1950s before the Unitary Plan tunnels were built, calibrated, and ready for experiments. They would eventually prove quite productive, but NACA researchers did not wait to begin high-speed experimentation. In the absence of new, more capable tunnels, researchers pursued competing alternatives, most of which had begun at Langley during the war. Earlier problems associated

---

36. Von Kármán, *Where We Stand*, p. 16. Von Kármán followed *Where We Stand* with *Towards New Horizons*, a multiauthor, multivolume work that analyzed the state of the art and gave R&D recommendations for the nation's Air Force. On Dryden's role, see Michael H. Gorn, *Hugh L. Dryden's Career in Aviation and Space*, Monographs in Aerospace History, no. 5 (Washington, DC: NASA SP-4500, 1996).
37. Roland, *Model Research*, pp. 211–221; NACA, "Policy for Operation of Unitary Wind Tunnels on Development and Test Problems of Industry," 6 May 1953, file 15861, NASA HRC. On the Air Force's push, see Neufeld, *Reflections on Research and Development in the United States Air Force*.

with the compressibility of air at high speeds had pushed researchers to consider the special conditions of aircraft as they approached and moved through the speed of sound. Of course, wind tunnels were the first recourse for investigators, but as velocities approached the speed of sound, the tunnels became choked. Shock waves and turbulence filled the test chambers, leaving behind a mess of unusable data. At velocities well above the speed of sound, shock waves abated and data became meaningful again. Researchers were left with a lacuna in the transonic regime.

At Langley, attempts to crack the transonic problem went in five directions, generally speaking. John Stack and his group approached the problem by designing a transonic research aircraft. Robert Gilruth's group used existing aircraft and merely placed aerodynamic shapes on the upper surface of an aircraft wing where local velocities approached and sometimes exceeded the speed of sound.[38] Another group used drop tests of vehicles and shapes from high-altitude bombers. Yet another group, underneath John Crowly, began using rockets. This latter method initiated the NACA's use of Wallops Island on the Virginia coast and was the beginning of the Pilotless Aircraft Research Division (PARD). Meanwhile, members of the wind tunnel research community continued to push and refine their own methods to better understand transonic and supersonic phenomena. Interestingly, these competing methodologies would remain extant, albeit under different people, for decades to come.

John Stack's vision of a high-speed research aircraft was slow to mature, but it eventually resulted in a series of aircraft, the first being the Bell XS-1, the first piloted aircraft to exceed the speed of sound (in controlled, level flight), with Charles "Chuck" Yeager at the controls. This would be the first in a long series of experimental vehicles (hence the X) that would continue into the NASA years. Stack's approach was not simply a successful technical alternative. It was a collaborative effort born of World War II and enabled by similar thinking within the Army Air Forces, notably Wright Field engineer Ezra Kotcher. The Army Air Forces and the Navy each procured high-speed research aircraft (the XS-1 in the case of the Army and the D-558 in the case of the Navy). Bell and Douglas designed and built the aircraft. The NACA provided technical design guidance and used the aircraft to explore the transonic regime. Though the aircraft would not begin flying until 1946 and 1947, respectively, the two programs established a division of labor among the participants and an infrastructure for exploiting experimental aircraft.[39]

38. Phillips describes this methodology in *Journey in Aeronautical Research*, pp. 91–97.
39. For information on the XS-1 and D-558 as well as flight testing at Dryden, see Richard P. Hallion, *Supersonic Flight: Breaking the Sound Barrier and Beyond* (London: Brassey's, 1997); and Hallion, *On the Frontier*.

(NASA image EC95-43116-6)

Figure 2.2. The staff of NACA Muroc in 1947, posing in front of the Bell XS-1 and B-29 mother ship.

The key development in the emergence of the high-speed flight program was a willingness on the part of the military to fund the design and construction of aircraft that would never see service. While the military had a history of building unique prototypes, such as the Douglas XB-19 (a very large prewar bomber) or the Bell XP-59 (the service's first jet aircraft), these earlier vehicles were funded with the expectation that they might evolve into production models. Although the XS-1 and D-558 programs were tinged with similar hopes (the D-558 especially), it soon became evident that these aircraft were esoteric research vehicles. Rather than building artifacts, the procurement process became an exercise in knowledge production. Not only did this legitimize, for the military, the category of experimental aircraft, but it took the military one step closer to the NACA's mission of fundamental knowledge production. Likewise, by piggybacking on military procurement and assisting in the design and production of actual vehicles, the NACA was also blurring the lines of its mission.

The establishment of the NACA's presence on the dry lakebeds of Muroc, California, began modestly with the arrival of a test pilot group and instrumentation experts from Langley. Langley had been flight-testing aircraft for years, examining new military aircraft, modifying existing aircraft, and, as mentioned above, using them to test airfoil shapes. Hand in hand with this, the Laboratory had developed a sophisticated instrumentation group that designed sensors and recording equipment. Both the Bell and Douglas test vehicles would fly with NACA instrumentation. By the time of Chuck Yeager's sound-barrier-breaking flight of 14 October 1947, there were 27 NACA personnel on hand from Langley's Flight Research and Instrument Research Divisions. In addition to instrumentation in the research aircraft, the NACA operated radar tracking and telemetry equipment. Yeager's flight was part of an initial Air Force testing program (thus handing the new military branch bragging rights). NACA testing, performed by NACA pilots, followed, and on 10 March 1948, NACA pilot Herb Hoover took the XS-1 supersonic, becoming the first civilian pilot to break the sound barrier.

In contrast to the record-breaking flights of the Air Force, the NACA's activities at Muroc emphasized methodical research (save Scott Crossfield's flight past Mach 2 in a Douglas Skyrocket).[40] Through the 1950s and a series of X aircraft, they examined the transonic aerodynamics issues that had prompted the development of the research airplane in the first place. They explored control and stability problems such as pitch-up and inertial coupling. They examined loading and vibration as well as the performance of a wide range of innovative configurations, including swept-wing, delta, variable sweep, and one vehicle with no horizontal stabilizer.[41] They validated the importance of all-moving stabilizers, thin wing sections, and vortex generators; as speeds pushed past Mach 2, they began to examine the problem of aerodynamic heating.[42]

Very quickly, the operation at Muroc established its own momentum. In 1947, still under the administrative control of Langley, the NACA renamed the group the Muroc Flight Test Unit. In 1948, Hugh Dryden formally placed control over research aircraft in the hands of Hartley A. Soulé at Headquarters. The Research Aircraft Program, as it was called, involved all of the Laboratories, but it concentrated most flight testing at Muroc. In 1949, the Muroc Unit became the High Speed Flight Research Station and earned its own budget. Finally, in 1954, it severed its relationship with Langley and became the High

40. Richard Hallion noted that Hugh Dryden initially forbade an NACA attempt at the Mach 2 record. "Dryden was well aware of the services' intention that records were to go to them, data to the NACA." Hallion, *On the Frontier*, p. 68.

41. For a useful technical review of variable sweep research, see Polhamus and Toll, "Research Related to Variable Sweep Aircraft Development."

42. In addition to Hallion's books, there is a very useful summary of the experimental aircraft: Dennis R. Jenkins, Tony Landis, and Jay Miller, *American X-Vehicles: An Inventory—X-1 to X-50*, Monographs in Aerospace History, no. 31 (Washington, DC: NASA SP-2003-4531, 2003).

(NASA image E-810)

Figure 2.3. Multiple-exposure photo of a Bell Aircraft Corporation X-5 at the High Speed Flight Station in September of 1952. The NACA and U.S. Air Force used the X-5 to study variable sweep wings from 1951 to 1955.

Speed Flight Station (HSFS), complete with its own administrative and support functions.[43]

The community of engineers, pilots, mathematicians, mechanics, and physicists at the HSFS grew into a highly skilled group. They became proficient at very dangerous operational tasks, such as rocket-plane fueling and airdrop launching. They learned to approach each test flight with meticulous attention to planning and safety. The HSFS also became an integral part of the Air Force's flight-test operations. Muroc, renamed Edwards Air Force Base in 1950, became the official Air Force Flight Test Center in 1951. Not only did the NACA depend on the Air Force and Navy for the procurement of X-vehicles, but they also received many prototype fighter, bomber, and transport aircraft for their own testing purposes. In return, the military received the NACA's constant support. Air Force test flights, for example, made use of the HSFS tracking and telemetry groups.

43. Hallion, *On the Frontier*, p. 31.

A measure of the strength of the HSFS is the fact that by the early 1950s, Langley engineers were well on the way to solving the original transonic wind tunnel problems that had prompted the concept of the research airplane in the first place. John Stack's group was again prominent in this search, and among their accomplishments was the development of model supports that reduced aerodynamic interference. A new breakthrough came from Ray H. Wright, a Langley physicist who found out how to implement a slotted tunnel (i.e., removing portions of the tunnel wall in order to eliminate troublesome transonic shock waves). The idea was not new; Italian émigré Antonio Ferri had worked on the problem in 1944–45. John Stack's group supported Wright's calculations, and by 1947, a test throat was in place in the 16-Foot High Speed Tunnel. With encouraging results, Stack pushed for full conversion of two tunnels, and by October of 1950, the 8-Foot High Speed Tunnel (HST) was in operation.[44]

The transonic tunnels began producing useful data fairly quickly. In what became one of Langley's signature discoveries, Richard T. Whitcomb came to the realization that large increases in transonic drag caused by wings could be offset by reducing an equivalent cross-sectional area in the fuselage. His breakthrough was in recognizing that the sharp increase in transonic drag occurred as a single shock wave that came from both the fuselage and the wing. Accordingly, his solution was to address how the entire aircraft generated this shock wave. He conducted the initial work in the 8-Foot HST and went on to develop a formal method of application by 1953. At the same time, Convair's new delta-winged supersonic interceptor, the F-102, was unable to break the sound barrier in level flight. By applying Whitcomb's method (the "transonic area rule") to the aircraft and creating a narrow-waist fuselage, designers enabled the aircraft to slip through the sound barrier, even in a climb.[45]

During and after World War II, the Aircraft Engine Research Laboratory (renamed the Lewis Laboratory in 1948) moved to make up for what was considered the NACA's most glaring failure (relative to Germany and England):

---

44. For an inside perspective on the transonic problem, see John Becker, *The High Speed Frontier: Case Histories of Four NACA Programs, 1920–1950* (Washington, DC: NASA SP-445, 1980). John Becker was an aeronautical engineer who joined the NACA in 1936 and remained at NASA to become both an influential researcher and laboratory leader.

45. Richard T. Whitcomb, interview by Robert G. Ferguson, telephone recording, 30 November 2005, transcript available at NASA HRC; Joseph Chambers, *Concept to Reality: Contributions of the NASA Langley Research Center to U.S. Civil Aircraft of the 1990s* (Washington, DC: NASA SP-4529, 2003), pp. 45–47; Hansen, *Engineer in Charge*, pp. 332–341. See Roland, *Model Research*, p. 287 and note 7.

(NASA image EL-2000-00280)

Figure 2.4. Area Rule Tests in the 8-Foot High Speed Tunnel, March 1957. Note the slots in the side of the tunnel, a critical Langley innovation that permitted the study of the transonic region.

the lack of turbine jet and rocket propulsion research.[46] The Lab built two supersonic tunnels late in the war. Work on rockets began without sanction from Headquarters after a member of the Laboratory, Walter Olsen, visited Caltech's Jet Propulsion Laboratory (JPL) and saw the work performed by Frank Malina's team. After the war, the Lab sought to distance itself from piston engine research. With the advent of the nuclear age, the Lab also developed plans for nuclear propulsion research, resulting in a joint laboratory with the Atomic Energy Commission, the Plum Brook facility, in the 1950s.[47]

---

46. See Virginia Dawson, "The American Turbojet Industry and British Competition: The Mediation Role of Government Research," pp. 128–130. On the history of the jet engine, see Edward W. Constant II, *The Origins of the Turbojet Revolution* (Baltimore: Johns Hopkins University Press, 1980).
47. On the history of the Lewis Lab (AERL), see Dawson, *Engines and Innovation.* On JPL, see Clayton R. Koppes, *JPL and the American Space Program: A History of the Jet Propulsion Laboratory* (New Haven, CT: Yale University Press, 1982); and Mark D. Bowles and Robert S. Arrighi, *NASA's Nuclear Frontier: The Plum Brook Research Reactor* (Washington, DC: NASA SP-4533, 2003).

(NASA image C-1955–37659)

Figure 2.5. Examination of a General Electric J-47 compressor section at the Lewis Lab in 1949.

The Lewis Lab established strengths in the aerodynamics and thermodynamics of turbine engines, fuels, and inlet/outlet behavior. It worked on practical matters such as icing and fire prevention. It added a Propulsions Systems Laboratory in 1952. In terms of jet engine research, the most valuable work took place in the Compressor Division. Focusing largely on axial flow engines, researchers attempted to find mathematical approaches that described compressor behavior (still a difficult task) and produced empirical engineering data that could be used by designers. High-altitude wind tunnels were especially important in this work. The Lewis Lab also concentrated on the problem of turbine blade cooling. As engineers sought higher engine speeds, which raised airflow temperature, they were pushing the limits of metallurgical capabilities. Lewis researchers conducted a wide range of tests on different solutions, including various materials, coolants, and design changes.[48]

Jet engine research also distinguished the Lewis Lab from the other Labs in its involvement with design and hardware issues. Though jet engines shared the

48. See the discussion of turbine cooling in Erik M. Conway, *High-Speed Dreams: NASA and the Technopolitics of Supersonic Transportation, 1945–1999* (Baltimore: Johns Hopkins University Press, 2005), pp. 31–32.

same theoretical and mathematical underpinnings as other aerodynamic structures, the fact that there were so many airfoils confined in a small space, many of them rotating and operating under very high temperatures, made the practical application of existing knowledge very difficult. As much as researchers sought to work first from theory, jet engines were so unpredictable and problematic that learning often began with the artifact. Lewis researchers got their hands dirty, becoming experts in component design on the way to understanding jet engine reliability and performance.[49]

The complexity of jet turbine aerodynamics, as well as the fact that each new turbine design represented a novel experiment, meant that the Lewis Lab had to work very closely with American manufacturers, dominated by General Electric, Westinghouse, Pratt & Whitney, and Curtiss-Wright. Much more so than Langley or Ames, Lewis's attachment to manufacturer's equipment made for a more restricted atmosphere. The protection of proprietary information was made policy in 1949.[50]

In rocketry, the Lewis Lab's efforts were minor in comparison to the research being conducted by the Army, Navy, and Air Force. Much of Lewis's work focused on propellants and was part of the Lab's general fuels research program (including, for example, jet engine fuels).[51] In spite of what would appear to be official resistance to rocket propulsion research at Lewis, as early as 1944, Jerome Hunsaker had proposed that the NACA move into guided missiles.[52]

Some of the most important work in rocketry actually took place at Langley. As noted above, researchers attempting to find their way out of the compressibility problem began using rockets to test aerodynamic shapes and models. In 1946, the Wallops launch facility formally became the Pilotless Aircraft Research Division (PARD). PARD did little to advance the NACA's work on rocket propulsion, but it was an important element in both high-speed, high-altitude research and electronics (including guidance and telemetry).[53]

On the West Coast, researchers at Ames were beginning to distinguish themselves from Langley's domination of aircraft-related work by pursuing more abstract and academic questions. Leading this charge was H. Julian "Harvey" Allen, a

---

49. See for example, Harold B. Finger, interview by Kevin M. Rusnak, tape recording, Chevy Chase, MD, 16 May 2002, p. 5 of transcript, NASA HRC. Virginia Dawson writes, "The machine itself was a teacher." Dawson, "The American Turbojet Industry and British Competition," p. 141.
50. Dawson, *Engines and Innovation*, p. 136.
51. Ibid., pp. 149–150.
52. Roland, *Model Research*, p. 252.
53. Ibid., pp. 253–254; Hansen, *Engineer in Charge*, pp. 267–270; Harold D. Wallace, Jr., *Wallops Station and the Creation of an American Space Program* (Washington, DC: NASA SP-4311, 1997).

former Langley researcher who had moved to California to start Ames. Through the 1950s and 1960s, Allen came to symbolize the kind of creative theoretician that the Ames community valued. In 1951, Allen and Al Eggers developed the blunt-body theory, which outlined how blunt-shaped nose cones could dissipate the heat of reentry by creating a shock wave in advance of the nose cone.[54] Such high-speed theoretical work was supported by new test instrumentation, notably Ames's supersonic free-flight tunnel (which fired objects into a supersonic stream) and Langley's hypersonic blowdown tunnels, which had begun operation in 1947.[55]

Between Sputnik and NASA

Prior to the launching of Sputnik on 4 October 1957, the NACA's commitment to spaceflight and rocketry represented the periphery of aeronautical research. High-speed research at Dryden, which by the late 1950s was building to the X-15 hypersonic program, had begun as an experiment into transonic and supersonic atmospheric vehicles. The work at Wallops Island was oriented largely toward aerodynamic experiments and the problems of guided missiles within the atmosphere. The work most closely associated with any future space program was, arguably, the theoretical and experimental work on reentry thermodynamics; and while as early as 1952, Robert Woods of Bell Aircraft urged the Committee on Aerodynamics to adopt the study of spaceflight, his proposal received only lukewarm approval. A review of the NACA's standing committees in January 1958 indicates the extent to which the organization was conceptually oriented toward the problems of atmospheric flight and of vehicles in particular:[56]

Standing Committees as of January 1958

- Committee on Aerodynamics
    - Subcommittee on Fluid Mechanics
    - Subcommittee on High-Speed Aerodynamics
    - Subcommittee on Aerodynamic Stability and Control
    - Subcommittee on Automatic Stabilization and Control
    - Subcommittee on Internal Flow

---

54. H. Julian Allen and A. J. Eggers, Jr., "A Study of the Motion and Aerodynamic Heating of Missiles Entering the Earth's Atmosphere at High Supersonic Speeds" (NACA TN-4047, 1957). This was originally published as a classified document under the same title but as Research Memorandum A53D28, 25 August 1953.
55. Hansen, *Engineer in Charge*, pp. 345–349. In a blowdown tunnel, air is released from a high-pressure vessel and passed through a test section. The Ames supersonic free-flight tunnel used blowdown in combination with a ballistic shot.
56. Roland, *Model Research*, p. 278, esp. footnote 33.

(NASA image A–23438)

Figure 2.6. H. Julian Allen stands at the observation window of the 8-by-7-foot test section of the Ames Unitary Plan Wind Tunnel, December 1957. A blunt shape, or body, sits inside the test section.

- Subcommittee on Low-Speed Aerodynamics
- Subcommittee on Seaplanes
- Subcommittee on Helicopters

- Committee on Power Plants for Aircraft
  - Subcommittee on Aircraft Fuels
  - Subcommittee on Combustion
  - Subcommittee on Lubrication and Wear
  - Subcommittee on Compressors and Turbines
  - Subcommittee on Engine Performance and Operation
  - Subcommittee on Power Plant Controls
  - Subcommittee on Power Plant Materials
  - Subcommittee on Rocket Engines
- Committee on Aircraft Construction
  - Subcommittee on Aircraft Structures
  - Subcommittee on Aircraft Loads
  - Subcommittee on Vibration and Flutter
  - Subcommittee on Aircraft Structural Materials
- Committee on Operating Problems
  - Subcommittee on Meteorological Problems
  - Subcommittee on Flight Safety
  - Subcommittee on Aircraft Noise
- Industry Consulting Committee[57]

Unlike its earlier experience with the turbojet, the NACA's reluctance to make spaceflight a priority did not reflect so much a lack of scientific creativity as an unwillingness to find political support for an area that was further afield from its traditional lines of research and one that was already claimed by the military. The Army, Navy, and Air Force each had their own missile and rocket programs; the NACA assisted these efforts but was in no position to intrude on such a highly contested area of weapons research.

From October 1957 to October of the following year, the NACA went from standing on the periphery to forming the backbone of a new organization for space and aeronautics, NASA. This was hardly a logical outcome, but, in the end, the NACA's civilian status, not its expertise in space or rocketry, gave the organization a distinct advantage over the Air Force's intercontinental ballistic missile (ICBM) program, over the Army's Redstone Arsenal headed by Wernher von Braun, and over the Naval Research Laboratory's Vanguard

57. NACA, "Functions and Responsibilities of Standing Committees and Subcommittees of the National Advisory Committee for Aeronautics," 1 January 1958, folder 15856, NASA HRC.

program. This was, largely, a top-down decision, but it was aided by the NACA's own strategic maneuvering.

Sputnik posed obvious challenges to the NACA's leadership. The organization had been experimenting on the doorstep to space and so could plausibly claim that it should strengthen these efforts. Likewise, whatever organization did receive the space mandate could quite possibly spell trouble for the NACA, subsuming its high-speed, high-altitude research and competing with it for R&D funding, not to mention stealing the limelight. The NACA had been burned before; its loss of leadership during World War II in areas of strategic military importance came at the cost of the Agency's reputation and may have encouraged the rise of competing functions within the Air Force.[58] Surely, with the political winds of Sputnik so easy to read, any reluctance to embrace space held familiar dangers, whereas support for space research held risks only for those who believed that the NACA was exclusively about aviation.

Jimmy Doolittle, who had taken over the NACA chairmanship from Jerome Hunsaker, and Hugh Dryden, the Director, began considering the space question only after much of Washington had already begun to wonder whether Eisenhower was doing enough to answer the Soviet challenge. On Capitol Hill, Senator Lyndon Johnson had begun his Inquiry into Satellite and Missile Programs, daily hearings that continued into January 1958, and maintained a steady drumbeat about the country's need to get into space quickly. Doolittle called a meeting in December at the Hotel Statler in Washington, DC, a month and a half after Sputnik, to discuss the question and to informally poll his researchers. Opinion was divided, but there was significant support from many to enter space, especially from the younger engineers and scientists. That same month, the agency also issued a report entitled "NACA Research into Space" that recast the organization's research as space oriented. While the report did highlight areas that were truly applicable to space, it overstated the organization's pre-Sputnik commitment to space. The effort, though somewhat disingenuous, was perhaps as much a rearguard action as it was an offensive maneuver.[59]

By January, however, the momentum was in the direction of making a full-fledged assault on space. On the 14th, the NACA issued a new report, "A National Research Program for Space Technology." It advocated research in the following areas:

---

58. See Roland, *Model Research*, p. 192, as well as the earlier discussion of the Air Force's AEDC.

59. NACA, "NACA Research into Space," December 1957 (unclassified extract published 10 February 1958), NACA Documents–1958, folder 15856, NASA HRC; Robert L. Rosholt, *An Administrative History of NASA, 1958–1963* (Washington, DC: NASA SP-4101, 1966), p. 34.

- Space Mechanics
- Space Environment
- Energy Sources
- Propulsion Systems
- Vehicle Configuration and Structure
- Materials
- Launch, Rendezvous, Reentry, and Recovery
- Communication, Navigation, and Guidance
- Space Biology
- Flight Simulation
- Measurement and Observation Techniques

More importantly, the report laid out the NACA's policy vision that the NACA would form the template and core of a civilian space science agency:

> The pattern to be followed is that already developed by the NACA and the military services. The NACA is an organization in being, already engaged in research applicable to the problems of space flight and having a great many of the special aerodynamic, propulsion, and structures facilities required, and qualified to take prompt advantage of the technical training and interest of scientists competent to help in the research on space technology. The membership of the NACA and its broadly based technical subcommittees includes people from both military and civilian agencies of the government, and representative scientific and engineering members from private life, thus assuring full cooperation with the military services, the scientific community and industry.[60]

On 16 January, the NACA passed a resolution that stated that space was part of the NACA's organic mission:

> Whereas, In the opinion of the Committee, the broad authority in its organic Act includes the investigation of problems relating to flight in all its aspects, outside of, or within the earth's atmosphere, of aircraft, missiles, satellites, and outer space projectiles and vehicles.[61]

---

60. NACA, "A National Research Program for Space Technology," 14 January 1958, folder 15856, NASA HRC.
61. NACA, "Resolution on the Subject of Space Flight," 16 January 1958, folder 15856, NASA HRC.

By the end of the month, Hugh Dryden was making public speeches on space policy and the role of the NACA: "In my opinion the goal of the [national space] program should be the development of manned satellites and the travel of man to the moon and nearby planets."[62]

The rhetorical argument from the NACA was that the military space program, which no one legitimately questioned, would engulf scientific space activity. As Dryden noted, the NACA provided a convenient way to balance competing interests: "There is another solution to the problem of how best to administer the national space-technology program, one which clearly recognizes the essential duality of our goals—the prompt and full exploitation of the potentials of flight into space for both scientific and military purposes. Actually, this solution is old and well-tested."[63] Dryden's vision was more restricted than what actually transpired, for he saw the NACA conducting scientific flights and, where called for, coordinating military, industry, and academic efforts.

Through the winter of 1958, the NACA built support for its plan. On 10 February, it published a multipronged research proposal for the exploration of space entitled "A Program for Expansion of NACA Research in Space Flight Technology with Estimates of the Staff and Facilities Required." In order to secure a position among the numerous competing space interests, the NACA established the Special Committee on Space Technology, which brought together experts primarily from outside the NACA. Only 3 of 16 members came from inside: Bob Gilruth of Langley, Abe Silverstein of Lewis, and Harvey Allen from Ames. If nothing else, the committee gave the appearance of the NACA standing at the center of scientific decision-making.[64] By March of 1958, the NACA had amended its standing committees to include missiles and spacecraft. For example, the Committee on Aerodynamics became

62. Hugh L. Dryden, "Space Technology and the NACA" (speech to the Institute of the Aeronautical Sciences, New York, NY, 27 January 1958), transcript, folder 15856, NASA HRC.

63. Ibid.

64. Membership included H. Guyford Stever (head), MIT; James A. Van Allen, Iowa State University; Wernher von Braun, Army Ballistic Missile Agency; Milton U. Clauser, Ramo-Wooldridge Corp.; James R. Dempsey, Convair; William H. Pickering, JPL; Hendrik W. Bode, Bell Telephone Labs; W. Randolph Lovelace, Lovelace Foundation; H. Julian Allen, NACA Ames; Col. Norman C. Appold, USAF; Dale R. Corson, Cornell; Robert R. Gilruth, NACA Langley; S. K. Hoffman, North American Aviation; Abraham Hyatt, U.S. Navy (USN); Louis N. Ridenour, Jr., Lockheed; and Abe Silverstein, NACA Lewis. NACA press release, "NACA Space Technology Committee Holds Organization Meeting," 14 February 1958, folder 15856, NASA HRC.

the Committee on Aircraft, Missile, and Spacecraft Aerodynamics; the underlying subcommittees remained nominatively unchanged.[65]

It is possible that the NACA would have been pulled into a space agency with or without the organization's lobbying. What is important to recognize, and what historians interested in the space story have not emphasized, is that Hugh Dryden, along with many of his aeronautics researchers, *chose* to take the NACA into space. For scientists and engineers at the time, the lines between atmospheric and space research were more blurred than they have since become. Space research was not necessarily a betrayal of one's past work, but, for many, a logical progression into a challenging and related arena (that was soon to receive a significant increase in research funding). This ambiguity between the two efforts would characterize the work and careers of many aeronautics personnel for the next few decades.

65. Rosholt, *An Administrative History of NASA, 1958–1963*, figures 2-1 and 2-2, pp. 25–26.

Chapter 3:

Creating NASA and the Space Race

In 1958, the Eisenhower administration decided to make the NACA the basis for a new national space agency. Seemingly, this is what Hugh Dryden and many of his researchers sought. In reality, the space program swallowed the NACA. Neither tragedy nor triumph, aeronautics was swept along on the coattails of the space program. Aeronautics, which for the past half century had symbolized cutting-edge research and daring technological exploits, was now lost in the roar of rockets and the allure of space exploration. Many aeronautics researchers took advantage of the opportunity posed by the transition to NASA, some jumping entirely into the new field, others moving back and forth across an often indeterminate boundary. Aeronautics, as a body of research, continued without significant disruption into the NASA era. While NASA managed large-scale research far differently than the NACA, notably in the weakening of the committee structure and in the increased reliance on contracting and project management, aeronautics research continued to rely on its relatively modest number of scientists and engineers at the four original Centers. Aeronautics funding, though a fraction of the space budget, grew slowly and continued to support what the NACA had formerly called "fundamental" research.

With space missions grabbing headlines and consuming the vast majority of NASA's budget, aeronautics was no longer a prominent political target. Even within NASA, aeronautics' visibility receded to the point that, by the early 1970s, Congress began to question whether the Agency was neglecting aeronautics. But even if aeronautics went unnoticed in the 1960s, it was not immune to NASA's larger structural shifts. Aeronautics began as a research-based bulwark in the face of a mission-oriented, publicity-driven organization.[1]

1. NASA was, of course, pursuing scientific and technological ends, but it was publicity-driven for precisely the same reasons that made Sputnik a cultural phenomenon and an embarrassment to Eisenhower and the United States.

(NASA image A-41727-6-4)

Figure 3.1. A fine example of the implementation of contemporary program management techniques. Merrill Mead, chief of Ames's Program and Resources Office, stands in a review room in the Ames headquarters building, October 1965.

A holdout from the NACA, its days were numbered. The transition toward mission-oriented aeronautical research began innocently as former aeronautics centers and researchers took on the project management of space missions and continued as Headquarters began to package aeronautical research as discrete, targeted, and closely managed projects. Over time, the original intention to maintain a sharp distinction between an operational space program and basic research began to give way to a belief that all of NASA should be run in the same fashion as the space program. Abetting this evolution was the porous boundary between aeronautics and space; on the one hand this gave researchers a wider range of topics to pursue, while on the other it opened the door for the spread of the space race's program management tools.

At NASA's inception, a number of aeronautical research programs lent themselves readily to the space program. Among them, the X-15 high-speed research aircraft represented both the climax of the NACA's experimental aircraft program and an important reusable vehicle for exploring flight at the edge of space. Likewise, wind-tunnel testing straddled both aeronautical and space research. Some space vehicles, after all, still had to negotiate the atmosphere. To this end, NASA's aeronautics researchers explored a number of reentry technologies, including lifting-body designs and deployable airfoils, such as the Rogallo wing. Such research attests to the eagerness of many aeronautical researchers to participate in the space program and a willingness to approach spaceflight as a natural evolution of atmospheric flight. More radically, the original Centers made moves toward owning pieces of the space exploration program.

Perhaps the best measure of the diminished importance of aeronautics to NASA's leadership was the design of the American supersonic transport (SST) program. In July 1960, when the matter came before NASA Administrator T. Keith Glennan, he asked that NASA not be given charge of the program. Instead, the Federal Aviation Administration (FAA), established in 1958 and led by an eager administrator, took on the project.[2] NASA provided research support to the FAA, but the space program was sufficiently demanding that NASA did not want the distraction of another major development program. The decision was entirely understandable even though NASA was the most logical and, arguably, most capable agency to lead the SST program. In what would become a recurring theme, the imperatives of the space program hobbled any significant top-level focus on, or commitment to, aeronautics.

The relationship between space and aeronautics has raised a persistent question over the years: whether aeronautics would have been better off going it alone. What is often forgotten is that even though the space program overwhelmed the NACA, the NACA model was under significant pressure to change in the late 1950s. It is quite possible that in becoming NASA, the NACA received a new lease and, for just a bit longer, could operate without burdensome oversight.

NACA to NASA: Riding a Tiger

Hugh Dryden's proclaimed vision for the NACA as a space agency was far more modest than what resulted in late 1958. Dryden's plan seemed to contemplate the least expensive and least offensive way to place the NACA at the center of

2. J. D. Hunley, ed. *The Birth of NASA: the Diary of T. Keith Glennan* (Washington, DC: NASA SP-4105, 1993), pp. 188–189; Mel Horwitch, *Clipped Wings: The American SST Conflict* (Cambridge, MA: MIT Press, 1982), chap. 2.

the nation's civilian space program. It steered well clear of the military, leaving untouched those programs dedicated to national security, as well as the military's investment in launch capability. As for civilian space science, it envisioned a large role for the National Science Foundation and the National Academies of Science. As in the NACA, Dryden's plan situated research decisions in a technocracy of scientific committees. The one remaining piece for the NACA was coordination, making sure that civilian space research made its way onto NACA-contracted launch vehicles. Dryden argued against the creation of new research centers to design and build spacecraft, stating, "It would be possible for NASA to build the organization and the facilities for such space vehicle and motor design and construction. But again, such action would be very costly and much additional time would be required."[3]

Eisenhower, and later President John F. Kennedy, did what Dryden and the NACA were too timid to suggest: the wholesale transfer of military launch vehicle programs to the new space agency. The process began in November 1957, soon after Sputnik's flight, when Eisenhower appointed James R. Killian, Jr., to be Special Assistant to the President for Science and Technology and created the President's Science Advisory Committee (PSAC). Killian was president of MIT and previously had chaired Eisenhower's Technological Capabilities Panel, which provided a roadmap for ICBM development and strategic reconnaissance technology (including military satellites). Even as Dryden's publicity machine geared up, Killian and the PSAC were coming to the conclusion that the NACA was the best candidate for a new space agency (announced on 4 February 1958). In spite of the NACA's effort to recast its work as space oriented, it was understood that the NACA was not as prepared as the military to enter space. However, the NACA offered Eisenhower three distinct advantages: it was civilian, and so sent a message to the world that the United States' intentions were peaceful; the NACA sidestepped interservice rivalry; and the NACA was politically weak, and thus would bend more easily to the administration's designs on space than, say, the more obdurate and quarrelsome branches of the military.[4]

---

3. Hugh Dryden, "A National Space Program for the United States," 26 April 1958, reprinted in Gorn, *Hugh L. Dryden's Career*, pp. 68–82. See also Rosholt, *An Administrative History of NASA*, pp. 8–35.
4. Walter A. McDougall, *...the Heavens and the Earth: A Political History of the Space Age* (New York: Basic Books, 1985), chap. 7. Ames aerodynamicist Vern Rossow recalled, "Dryden was one of us, not political. He was fine when the organization was small; he could go to Congress and explain the technical details. But he wasn't appropriate for a large organization with a larger budget and larger political challenges." Vernon J. Rossow, oral interview by Robert G. Ferguson, Ames Research Center, 13 and 18 January 2005, transcript at NASA HRC.

Dryden could count Killian and a number of the PSAC members as friends, but this did little to further Dryden's plans. The new space agency was not to be an NACA with a space mission but a comprehensive space agency that enveloped the NACA. With the country already second in the publicity race with the Soviets, there was no time for the NACA to learn how to build its own launchers or to develop the kinds of program management skills that had become stock-in-trade elsewhere since World War II. The NACA would need more than just access to the military's space resources—military programs would script the organization's new mode of operation. NASA was to be a much larger organization that made use of extensive subcontracting arrangements and exchanged its committee structure for a hierarchy answering to the President.[5]

Eisenhower directed members of the PSAC, Bureau of the Budget, Rockefeller Commission on Government Organization, and NACA to draw up the necessary legislation. The NACA's leadership fought for, and ultimately lost the battle over, the committee structure, over which there had been long-standing objections, though they were given the trivial concession of a toothless advisory board.[6] In April 1958, Eisenhower sent the National Aeronautics and Space Act to Congress. The Act gave the military control over weapons systems and reconnaissance satellites, but the vast remainder went to the civilian space agency. Congress debated the bill from April until July, when Eisenhower and Lyndon Johnson met to resolve differences; the bill passed at the end of the month. The new agency came into being on 1 October 1958 with Eisenhower choosing T. Keith Glennan, president of Case Institute of Technology, as the Agency's first Administrator. Hugh Dryden, considered a relic of the more conservative NACA, was the first Deputy Administrator.[7]

The irony of the NACA's bold, opportunistic gamble for space is that aeronautics quickly disappeared. It is not that aeronautics research ceased, but that, on NASA's first day of operation, aeronautics had vanished from the foreground. Space was NASA's mandate. What was to become of aeronautics, and what were the great challenges facing this five-decade-old enterprise? As historian Alex Roland wrote in the last sentence of his history of the NACA, "...the NACA was laid to rest because it had accomplished what it set out to do."[8] For the next decade, there would be many questions and debates about

5. Rosholt, *An Administrative History of NASA, 1958–1963.* Also helpful is Pam Mack, "NASA and the Scientific Community: NASA-PSAC Interactions in the early 1960s," 4 May 1978, folder 123000, NASA HRC.
6. Roland, *Model Research*, pp. 294–296.
7. McDougall, *...the Heavens and the Earth: A Political History of the Space Age.*
8. Roland, *Model Research*, p. 303.

NASA, but aeronautics would hardly register. Not until the end of the Apollo program would aeronautics reenter the fray, and even then, it could hardly be called an equal of the space program.

Within NASA, aeronautics did not constitute its own budgetary or research area but was initially grouped in the Office of Aeronautical and Space Research (OASP). The OASP took control of the existing four NACA Centers and was to operate as the basic research component of the Agency. The Office of Space Flight Development took in Goddard Space Flight Center in Beltsville, Maryland; JPL; the Wallops Flight Facility in Virginia; and Cape Canaveral, Florida (which later became Kennedy Space Center), and was to be the developmental and operational side for space activities. In early 1961, the OASP core became the Office of Advanced Research Programs, and later that year, after James Webb replaced Glennan as Administrator, it changed names again to become the Office of Advanced Research and Technology (OART). The 1961 reorganization placed all of the Centers underneath the Associate Administrator, but by late 1963, they were again attached to specific programs, and the original four Centers were back with OART. Elsewhere in NASA, the Centers at Huntsville and Houston, as well as Kennedy Space Center, operated under the Office of Manned Space Flight, while Wallops, Goddard, and JPL operated under the Office of Space Science and Applications. Like the aeronautics program, the latter Centers (which had robotic exploration and astronomy programs) were dwarfed by NASA's human space program.[9]

The original intention behind the creation of OASP and OART was that these programs would serve as the basic research core for the rest of NASA and would perform only limited developmental work. P. M. Lovell, the Assistant to the Director of Aeronautical and Space Research argued that only the developmental centers should have defined research areas, and that "the mission of the research centers [Langley, Lewis, Ames, and the Flight Research Center] can only be defined in the broadest possible terms and that a statement of the current research being done will only be applicable for a very short period of time. For example, the mission of the research centers as to research may be defined

9. For comparative data on the human space program budget versus all other NASA initiatives, consult the *NASA Historical Data Book* series (NASA SP-4012): Jane Van Nimmen and Leonard C. Bruno with Robert L. Rosholt, *NASA Historical Data Book, Volume I, NASA Resources 1958–1968* (Washington, DC: NASA, 1988), table 4-22; Ihor Gawdiak with Helen Fedor, *NASA Historical Data Book, Volume IV, NASA Resources 1969–1978* (Washington, DC: NASA, 1994), table 4-21; and Judy A. Rumerman, *NASA Historical Data Book, Volume VI, NASA Space Applications, Aeronautics and Space Research and Technology, Tracking and Data Acquisition/Support Operations, Commercial Programs, and Resources 1979–1988* (Washington, DC: NASA, 2000), table 8-12.

as merely to study the problems of flight within and outside the atmosphere." The tenor of the time was that the Centers informed Headquarters, which made certain that there was no duplication of effort but did not control the research. Lovell also wrote, "[I]t should be realized by all NASA management personnel that the research of the research centers is quite different from the research of the development centers."[10]

Briefly, Langley, Ames, Lewis, and the Flight Research Center (FRC) were known as the "Advanced Research Centers," charged with "providing information required for the accomplishment of the Nation's space and aeronautics goals not attainable with current knowledge." The new goals included the following:

1. Provide a broad base of research which allows the development of technology necessary for manned and unmanned exploration of space, and for the advancement of aeronautical vehicles;
2. Generate new and advanced concepts for future space and aeronautical vehicles and their expected missions; and
3. Provide the entire NASA organization, the Department of Defense, and others with needed research assistance.[11]

The four Centers were given specific duties:

- **Langley:** materials, structures, magnetogasdynamics, aerodynamic heating, guidance, and navigation.
- **Ames:** hypersonic aerodynamics; orientation, stabilization, and control; and space environmental physics.
- **Lewis:** advanced propulsion systems, power-generation systems, and basic materials research.
- **FRC:** flight mechanics, operation and environment, and aerodynamic and structural effects.[12]

The initial division of labor between the research centers and the mission centers did not survive over the long term. Making this kind of organizational distinction simply did not work because it disregarded the degree to which "fundamental" research, as practiced by the NACA, relied on development work. Likewise, the "developmental" side of NASA relied on a great deal of its own experimental research. Additionally, between the prestige attached to basic research and the competition for resources, there was little incentive for programs outside of OART to willingly hand over experimental or

10. P. M. Lovell to Mr. Rhode, "Comments on Material from Dr. Dryden," 16 June 1959, OAST Correspondence 1959–1962, folder 18269, NASA HRC.
11. NASA, "Fiscal Year 1962 Estimates, Support of NASA Plant, Advanced Research Centers," undated, OAST Correspondence 1959–1962, folder 18269, NASA HRC.
12. Ibid. (summarized, not directly quoted).

knowledge-producing activities. Over the long run, the failing distinction between basic and developmental work meant that OART and its centers had to compete on the basis of what they did rather than how they did it.

In the first few years of NASA, aeronautics began a subtle transition. On the one hand, research activities at the original Centers continued much as they had in the 1950s. They were, for the most part, under the same leadership. Researchers expected their work to continue as before, though with a new mandate that broadened the scope of their work. However, at Headquarters, the independence of the old Centers stood in contrast to the more centralized and highly coordinated human space program. Even though aeronautics was a lower priority, there was clear discomfort in not knowing exactly what the aeronautics people were doing. In the past, Headquarters had merely signed off on research topics and budgets sent from the Centers, a task that was both infrequent and minimally informative. Ira Abbot, the Director of the Advanced Research Programs Office in 1960, discussed with the Center Directors the need for more thorough reporting in order to coordinate research across the Agency. "The most serious deficiency of the [Administrator's Progress Report] is that it does not present adequate information concerning the general research activities of the centers."[13]

Unable to ascertain what was going on at the Centers, Abbot in particular and Headquarters in general had little ability to exercise the same kind of control exercised over the space program. Through NASA's first few years, the Agency was playing catch-up with the aeronautical work at the Centers. In 1960, NASA created the Research Program Analysis System (RPAS), with a first step being the cataloging of in-house and contracted research.[14] The following year brought the OART reorganization, as well as a new Office of the Director of Technical Program Coordination. The latter was to coordinate OART's research with the needs of the other directorates, make long-range plans for where OART's research should go (through parametric studies of space systems), and coordinate communication between NASA groups, as well as NASA and outside agencies.[15] At the same time, OART was broken

13. Ira H. Abbot to Langley, Ames, Lewis, and Flight, "Administrators Progress Report," 9 September [likely 1960], OAST Correspondence 1959–1962, folder 18269, NASA HRC.
14. Ira H. Abbot to Directors of Space Flight Programs, Launch Vehicle Programs, and Life Sciences Programs, "Formal Coordination of NASA research activities," 1 December 1960, OAST Correspondence 1959–1962, folder 18269, NASA HRC. See also Ira H. Abbott to Langley, Ames, Lewis, and Flight, "Documentation of the NASA Research Programs," 26 April 1961, OAST Correspondence 1959–1962, folder 18269, NASA HRC.
15. "Technical Program Coordination," 7 November 1961, OAST Correspondence 1959–1962, folder 18269, NASA HRC.

into technical directorates: research (fluid physics, materials, electromagnetics, and applied magnetics), nuclear systems, propulsion and power generation, aeronautics, electronics and guidance, and space vehicles. Each of the technical directorates was to perform long- and short-range program planning, evaluate and recommend programs to the OART Director, establish technical guidelines and schedules, serve as the Headquarters point of contact for research persons at the Centers, monitor progress, formulate program budget guidelines, and evaluate personal performance.[16] Finally, Headquarters created an Office of Program Review and Resources Management that would establish management control systems, program analysis, and facilities planning.[17]

Change at the Centers

For researchers at the Centers, the experience of becoming a space agency was mixed. For those who were only minimally engaged in space activities, the early years of NASA were traumatic. Historian James Hansen described widespread resentment and emotional resignation about the decline of aeronautics at Langley, leading most notably to the departure of one of Langley's stars (and Director of Aeronautics) in 1962, John Stack.[18] Some aeronautics researchers found themselves asked to manage space projects when they would have preferred to continue their aeronautical investigations.[19] For others, space presented opportunities for entirely new research and career paths. Most certainly, Sputnik reoriented researchers toward space, and work that had been classified as aeronautics was reclassified as space related. For example, table 3.1 provides the budget for aeronautics within the OARP from 1955 to 1960. Aeronautics appears to suffer a decline from 89 percent of funding in 1955 to an estimated low of 27 percent in 1962. While Sputnik gave space research a considerable boost, some aeronautics research simply *became* space research. Langley engineer W. Hewitt Phillips writes that upon becoming NASA, "every group immediately considered how its knowledge and expertise could be applied to aid the space effort."[20]

---

16. "Statement of General Responsibilities," 7 November 1961, OAST Correspondence 1959–1962, folder 18269, NASA HRC.
17. "Program Review and Resources Management," 7 November 1962, OAST Correspondence 1959–1962, folder 18269, NASA HRC.
18. Hansen, *Spaceflight Revolution*, pp. 97–102.
19. James R. Hansen, *The Bird Is on the Wing*, Centennial of Flight Series, no. 6, ed. Roger D. Launius (College Station: Texas A&M University Press, 2004), pp. 146–147.
20. Phillips, *Journey in Aeronautical Research*, p. 168.

Table 3.1. Office of Advanced Research Programs Budget
(S&E and R&D; not including C&E)

Fiscal Year	Total OARP	Total in Aeronautics	Percent
1955	$54,240,000	$46,494,000	89
1956	$60,135,000	$50,513,000	84
1957	$64,176,500	$49,416,000	77
1958	$71,000,000	$44,730,000	63
1959	$77,750,000	$33,432,000	43
1960	$85,269,590	$27,286,000	32
1962 *(estimated)*	$112,038,000	$30,021,000	27

Source: OARP budget table, prepared 13 January 1961, OAST Correspondence 1959–1962, folder 18269, NASA HRC.

Table 3.2. Distribution of Research Professionals, Office of Advanced Research Programs, 1960

Area	Number	Percent
Structures	199	9
Materials	197	9
Flight Mechanics	307	14
Hypersonic Aerodynamics	268	12
Nuclear Rockets	162	7
Chemical Rockets	238	11
Power Generation	32	1
Electrical Rockets	73	3
Instrumentation and Data Acquisition	88	4
Operation and Environment	184	8
Supersonic Aerodynamics	276	12
Air-Breathing Engines	89	4
Subsonic Aerodynamics	148	7
Total	*2,261*	

Source: Director Advanced Research Projects to Director Business Administration, "FY 1962 Detailed Budget Submission," 31 August 1960, OAST Correspondence 1959–1962, folder 18269, NASA HRC.

Table 3.3. Research Center Funding Allocation by Program

Aeronautics Program	Percent
VTOL/STOL Aircraft	3.0
Subsonic Aircraft	1.4
Supersonic Aircraft	7.6
Hypersonic Aircraft	6.9
Non-Vehicle Oriented	1.1
Subtotal	*20.0*

Space Program	Percent
Launch Vehicles	19.4
Manned Earth Satellites	22.5
Lunar and Interplanetary Vehicles	24.9
Unmanned Earth Satellites	6.2
Ballistic Missiles	1.7
Maneuverable Missiles	3.7
Non-Vehicle Oriented	1.6
Subtotal	*80.0*

Source: "Distribution of Research Center Budget," 7 March 1961, OAST Correspondence 1959–1962, folder 18269, NASA HRC.

Table 3.4. Research Center Funding Allocation by Area

Area	Percent
Hypersonic Aerodynamics	18.4
Supersonic Aerodynamics	12.2
Subsonic and Transonic Aerodynamics	9.8
Chemical Rocket Propulsion	9.5
Structures	8.0
Fluid Mechanics	7.7
Nuclear Rocket Propulsion	5.7
Space Environmental Physics	4.7
Materials	4.2
Guidance and Navigation	4.2
Material Applications	3.8
Data Acquisition and Transmission	3.6
Electrical Propulsion	2.7
Control and Stabilization	2.7
Electric Power Generation	1.5
Operating Problems	0.7
Air-Breathing Engine Propulsion	0.6

Source: "Research Center Effort By Area," 7 March 1961, OAST Correspondence 1959–1962, folder 18269, NASA HRC.

Table 3.3 shows an 80/20 split in funding between space and aeronautics activities at the OARP/OAST Centers in 1960. At the same time, table 3.4 shows that the various subdisciplines of aerodynamics received over 40 percent of total funding. Additionally, aerodynamicists constituted a third of OARP's research personnel (table 3.2). Some of the discrepancy arises from the fact that work in space launchers and vehicles involved a greater degree of outside subcontracting, so increased space funding did not translate into a proportional increase in NASA space research personnel. The mingling of aeronautics and space within the same organization and the porous boundaries between the activities did, however, allow the Centers some freedom in how they categorized their work.

Continuity in research was accompanied by continuity in leadership. At Ames, Smith DeFrance, who had been running the facility since 1940 and was himself a Langley veteran, continued to serve until 1965 as Center Director. H. Julian Allen, who had accompanied DeFrance from Langley, succeeded DeFrance. Not until 1969 did Ames bring in an outsider, Hans Mark. Both DeFrance and Allen managed to protect the laboratory, serving as gatekeepers between the Center and Headquarters. For researchers, the budgetary process and the politics of NASA were largely opaque. As Jack Boyd, a researcher and Associate Director at Ames noted, "As engineers, we never knew anything about dollars. We had no budgets; we just ran what we wanted to run, designed what we wanted to run, and Smitty [DeFrance] protected the boundaries of the Center completely."[21]

Where some researchers felt threatened by the switch to NASA, many others embraced it.[22] Scientists and engineers who had spent their lives working on aeronautical problems retooled for the new tasks. Vic Peterson, an Ames aerodynamicist, recalled, "A lot of us were interested in gravitating toward the space area." For Peterson, this led him to begin thinking about how vehicles might fly elsewhere, such as Mars's carbon dioxide—nitrogen atmosphere. This interest led to pumping carbon dioxide and nitrogen into the wind tunnels to perform tests. "So we were gung-ho about it, actually, and thought that we could have a little of both. We could have our aeronautics and apply it to flying in atmospheres of other planets."[23] Other individuals moved more squarely into space activities. Charlie Hall, who had been the Branch Chief in the Ames 6-by-6-Foot Wind Tunnel, eventually became the project manager for the Pioneer spacecraft program, which was run out of Ames.[24]

Just as individual researchers wandered afield of aeronautics, so did the Centers. Of the four original Centers, Lewis was the most aggressive in moving into space propulsion and power generation. By 1960, air-breathing engines

21. John W. Boyd, interview by Robert G. Ferguson, tape recording, 14 January 2005, Ames Research Center, NASA HRC.

22. Elizabeth A. Muenger quotes Ames engineer Seth Anderson: "...what about aeronautics? It shouldn't stop; there were still plenty of things to do, and we were still interested in doing aeronautics." *Searching the Horizon*, p. 96.

23. Victor L. Peterson, oral interview by Robert G. Ferguson, Los Altos, CA, 17 January 2005, NASA HRC.

24. Boyd interview, HRC.

accounted for a mere 7 percent of the Center's work.[25] For Lewis, the shift to space-related work represented a much less excruciating reorientation than for those engineers focused on aircraft. After Lewis, Ames took the greatest risks in branching out. In part, Ames was in danger of being neither fish nor fowl. It was not a space center, and Langley retained its preeminent position in aircraft and applied research. Langley had a structures laboratory that Ames lacked; Langley also continued to oversee the development of high-speed aircraft. Ames sought out space-related areas such as navigation and the life sciences. Harvey Allen encouraged DeFrance to move Ames into biology as the Center already had experience in human factors research.[26] These efforts eventually led to a thriving astrobiology program. Ames also pursued flight biomedicine, which, along with its future thrust into computing, set it on a course to building a strong simulation program.[27]

Langley also sought to branch out beyond aeronautics. As queen bee of aeronautical research, it had the most at stake in the transition to NASA. Initially, Langley engineers such as Max Faget dominated the Space Task Group that would form the kernel of the NACA's contribution to the space program. Unfortunately, Langley lost its early space group to the new space mission centers, such as Johnson and Marshall.

Even if aeronautics researchers did not avail themselves of the new opportunities presented in the NASA era, the tenor of existing research was beginning to change. The NACA had always been a knowledge producer, even in those instances where it found itself building an aircraft or rocket-plane. The end product was always a technical paper or advice distributed to the military or industry. In the NASA era, knowledge production gave way to the completion of big projects. The size of research expanded, and, increasingly, it involved teams of researchers working in conjunction with subcontractors. With big projects came big money and big management. This was in marked contrast

25. Dawson, *Engines and Innovation*, p. 179.
26. Boyd interview. Though exobiology was a long way from Ames's original strengths, the addition of an academically oriented department provided intellectual stimulation and contributed to Ames's self-perception as the most cerebral of NASA's aeronautics Centers. Aerodynamicist Vern Rossow recalled frequenting the astrobiology group's lectures, and while there was no direct application of the life sciences on Vern's work, he found it inspirational nonetheless. Rossow oral interview. Vic Peterson described Ames as being in a hotbed of technology, academics, and entrepreneurship and being the most educated of the Centers. Victor L. Peterson, oral interview by Robert G. Ferguson, 17 January 2005 in Los Altos, California, NASA HRC.
27. Muenger writes that flight simulation research was one of Ames's responses to perceived threats from the space program. *A History of Ames Research Center*, p. 96.

to the lone researcher working on a topic of personal interest on bootstrapped wind tunnel time. With project management came more accountability for where money was spent, though salaries and infrastructure (such as wind tunnels) continued to be funded as recurring expenses rather than part of the research budget into the 1980s.

Subtler were the changes that began to occur with in-house publications and in-house presentations. Tradition was that researchers published their work in NACA technical memos, notes, and reports. When researchers gave presentations, it was common for branch chiefs and Center administrators to attend. With the dawn of NASA, in-house publications began to wane. Perhaps as part of the influence of new, academically oriented disciplines, as Jack Boyd speculated, aeronautics researchers sent their publications to external peer-reviewed journals. Internal peer-review committees fell by the wayside, as did the leadership's ability to stay abreast of all their researchers' activities.[28]

X-15 on the Near Side of Space

The X-15 program began in the NACA years and continued through NASA's first decade. A hypersonic rocket-plane, the X-15 was the apotheosis of the high-speed research plane approach that had begun in the 1940s with the XS-1. The X-15 was not the last of NASA's experimental vehicles, but it was most certainly the end of the line of development that saw the open-ended construction of supersonic vehicles. The X-15 was meant to go fast and high, and whatever other fruit it bore was serendipitous.

The X-15 was arguably NASA's last high-profile aeronautics program (insofar as the public was even aware of such research). The X-15 is, for example, the only NASA aircraft to occupy a place in the National Air and Space Museum's Milestones of Flight Gallery (the Bell X-1 being an NACA aircraft). Certainly, the X-15 benefited from being a tangible artifact in a way that wind tunnel research is not, but as a daring experiment that took pilots higher and faster than ever before, it also shared in the astronaut mythology of the space program. Indeed, 5 of the 12 pilots who flew the three aircraft earned their astronaut wings. The X-15 was NASA's last nonspace program to create American heroes. As an icon of American technological accomplishment and as a tribute to the cultural fascination with speed, the X-15's symbolic status arguably trumps, at least from a historical perspective, the aircraft's scientific value.

Prior to the creation of the X-15 program in 1954, the NACA had flown as fast as two times the speed of sound and had reached altitudes of up to

28. Boyd interview.

90,000 feet. With experiments already under way during the mid-1950s, pilots would pass Mach 3 and 120,000 feet. The NACA, military, and industry understood that going faster and higher entailed significant problems with aerodynamic heating, control, and pilot safety. Though temperatures at high altitudes are well below freezing, the simple action of air rubbing against the aircraft skin at high speed raises temperatures enough to potentially melt aluminum and soften most other available alloys. As early as 1951, researchers at North American Aviation's Aerophysics Laboratory expressed concern about aerodynamic heating and asked that the NACA address the challenge.[29] Such heating was not readily modeled in wind tunnels, which are better equipped to demonstrate streamlines, turbulence, and shock waves. High-speed flight went hand in hand with flying at higher altitudes, where the atmosphere was less dense. This, however, presented novel control problems because control surfaces had little to "push" against. Finally, the pilot required new pressurized clothing and safety devices in order to survive the extremes of temperature, pressure, and acceleration. For all practical purposes, these pilots would be short-duration astronauts.

With the momentum of the early X-vehicles, it is easy to see the X-15 as the next logical development. For the most part, it shared the same organizational features of earlier programs: this was a joint NACA–Air Force–Navy program in which the NACA did the brainwork, the Air Force funded and administered the program, and the Navy contributed the remaining costs. It would operate out of the Edwards Air Force Base (AFB)/NACA High Speed Flight Station area (though with a much expanded flight and tracking area due to the aircraft's velocity). It would be an air-dropped rocket-plane manufactured by a private contractor. Where the X-15 defied expectations was its transition from a vision of hypersonic aircraft to a vision of reusable spacecraft. The X-15 did not merely span the NACA's institutional shift to NASA, it pivoted between two very different research paradigms. Though the NACA had only lukewarm interest in space in the early 1950s, the X-15's designers created a robust research platform that, among other things, had the ability to fly to extreme altitudes. In the wake of Sputnik, the X-15 provided a way for the NACA to argue that it was, in fact, headed to space. Rolled out from North American's factories on 15 October 1958, it was NASA's first tangible spacecraft. Thus, against most people's expectations, the rocket-plane's flexibility allowed the X-15 to take on an entirely different

29. Edwin Hartman to Director, NACA, 16 July 1951, "Visits to the North American Aerophysics Laboratory at Downey, California, 22 June and 3 July 1951," Edwin Hartman Memorandums, Langley Research Center Historical Archive.

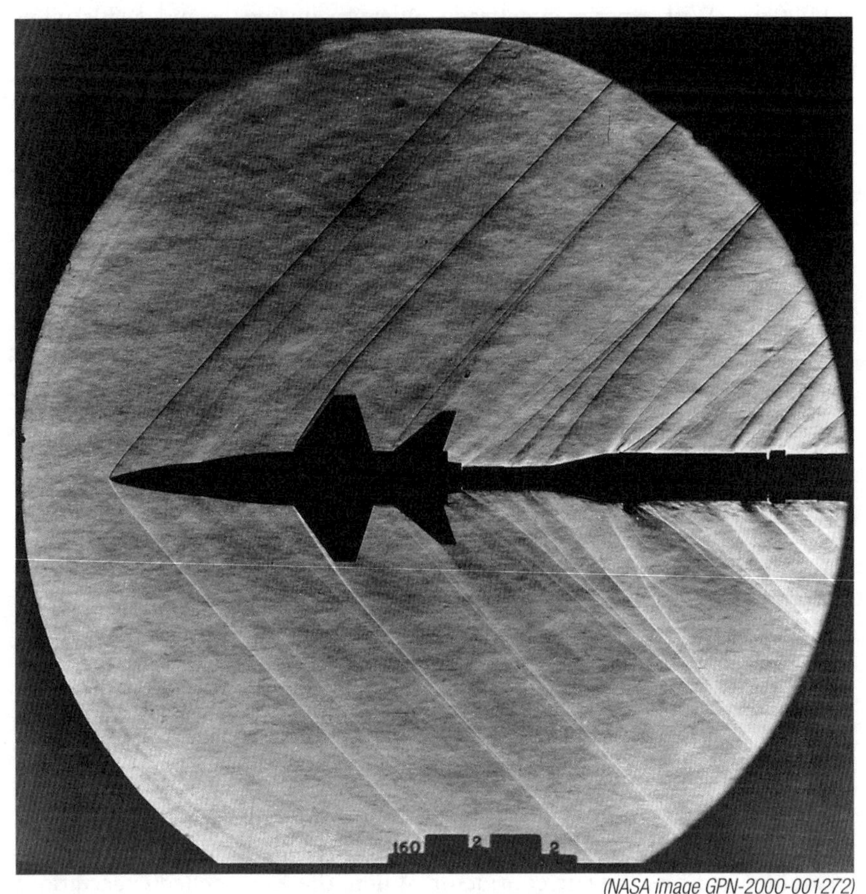

(NASA image GPN-2000-001272)

Figure 3.2. Scale model of an X-15 in the Langley 4-by-4-Foot Supersonic Pressure Tunnel in the early 1960s. The lines angling away from the model are shock waves.

meaning in the 1960s. Far from being a logical technical evolution, the X-15 research program reflected NASA's institutional bent and the nation's broader Cold War struggle.

Hypersonic research had its promoters in the NACA as early as 1952, and there had been conceptual discussions of such an aircraft in wartime Germany. Not until 1954 did the idea receive an official blessing from the NACA, which invited proposals from the Centers.[30] However, Langley was the likely

30. Hallion, *On the Frontier*, pp. 106–107.

candidate with its coterie of experienced rocket-plane alumni.[31] John Becker headed Langley's design team, a group that included Norris Dow, Maxim Faget, Thomas Toll, and James Whitten. In arriving at a hypersonic aircraft design, Becker's team added a twist: not only would their aircraft fly fast, it could also fly at a very steep angle and thus pierce the edges of space. This latter element was hardly in keeping with the plane's intent, but the NACA chose the Langley team and retained the feature.

By the end of 1954, the NACA had set out the X-15's general design goals: the exploration of aerodynamic heating, stability, and control problems of hypersonic, high-altitude flight; physiological factors of hypersonic flight; and the potential for including an observer (a feature that fell by the wayside).[32] In May 1955, Bell, Douglas, North American, and Republic submitted design proposals to the Air Force. By late summer, the NACA, Air Force, and Navy had selected North American as the preferred designer and manufacturer, and by late fall, contracts and funds were forthcoming. Though Becker's group had defined many of the aircraft's critical parameters, North American would have to solve many of the specific technical issues. For the next three years and under the leadership of former NACA research pilot Scott Crossfield, North American employees would work closely with their NACA counterparts. The company would also have recourse to the NACA's hypersonic tunnels, the 11-inch at Langley and Ames's free-flight facility.

Where shock waves and the transonic regime were considered the major challenges of the Bell XS-1, aerodynamic heating was the major challenge of the X-15. The aircraft would routinely experience temperatures of 1,200 degrees Fahrenheit, though without accurate models for their hypersonic wind tunnels, Langley's engineers had to design conservatively and anticipated temperatures of up to 2,000 degrees. They considered two approaches to keeping the aircraft from melting: insulating the structure with different layers of materials and building the outer shell of a material that simply absorbed the heat. They chose the latter and fabricated the outer structure of the aircraft with a nickel chromium alloy called Inconel X. For much of the inner structure, including the complex truss and corrugated sheet wings, the designers used titanium. Making the structure strong was less difficult than anticipating the results of

31. The history of the X-15 program is covered in a number of secondary sources, including Hallion's *On the Frontier* and Dennis R. Jenkins, *Hypersonics Before the Shuttle: A Concise History of the X-15 Research Airplane*, Monographs in Aerospace History, no. 18 (Washington, DC: NASA SP-4518, 2000).

32. See "Preliminary Outline for High-Altitude, High-Speed Research Airplane," 15 October 1954, reproduced in appendix 3 of Jenkins, *Hypersonics Before the Shuttle*.

differential heating (e.g., the bottom of the vehicle was far hotter than the top), different rates of expansion (that led to distortions in the vehicle's shape), and local hot spots (that resulted from any kind of airflow perturbation). Aerodynamic heating was also a function of the aircraft's flight profile (e.g., the aircraft's velocity, attitude, and air density). Designers had to make provisions for keeping the aircraft at the proper attitude and for slowing the vehicle as it reentered the atmosphere.

To solve the stability and control problems, designers had to tackle two distinct challenges: how to design control surfaces that operated at high speeds and how to design control systems that operated out of the atmosphere. To address the former, North American engineers made use of a large, all-moving wedge for the vertical tail, speed brakes at the rear of the aircraft, and differentially operated horizontal slabs (that eliminated the need for ailerons). For control above the atmosphere, designers placed small rockets and reaction jets in the nose and wings. The pilot had to switch between the two systems while in flight (using different joysticks) since coupling the aerodynamic surfaces and reaction jets would have required knowledge that the X-15 was attempting to learn (e.g., how to modulate both systems through an entire flight profile), as well as computer systems that were not yet up to the task. Only late in the X-15 program and with the last of three aircraft did NASA implement a control system that sought to automatically integrate the two systems.

The X-15 was very much a learning device for design and manufacturing. Systems that operated well at supersonic speeds and below 100,000 feet became faulty at lower pressures and higher temperatures. The NACA and North American had to invent solutions all along the way. Rollout of the first aircraft, as noted above, took place in October 1958. It was the rocket engine, however, that presented considerable delays. Interestingly, for all the novelty of the engine, the XLR99 built by Reaction Motors, propulsion was not considered one of the main goals of the program. This highlighted the importance that the NACA attached to aerodynamics, structures, and atmospheric flight over spaceflight. The XLR99 was fueled by liquid oxygen and anhydrous ammonia, and it was the first throttleable and restartable human-rated rocket engine. The engine would ultimately provide 57,000 pounds of thrust (for a vehicle that weighed only 30,000 pounds). It was 1959 before engines were available for initial testing.

A full five years after the NACA initiated the X-15 project, North American began its own test program to ensure that the vehicle met specifications. With Scott Crossfield at the controls and carried underneath the wing of an early-model B-52, the first drop test took place in June 1959, and the first powered test in September. North American handed over the first of the X-15s in February 1960. Though the X-15's flights were usually under a dozen minutes, it traveled such long distances that NASA created a new test range. Called the "High Range," it stretched from Utah to California and incorporated three radar and telemetry stations that could seamlessly follow each test. Early flights involved working out the bugs in the X-15's novel systems, especially the rocket engine, controls systems, and auxiliary power unit. By 1961, X-15 pilots were beginning to make flights above Mach 4 and reaching skin temperatures of 500 degrees Fahrenheit. By the end of the year, the pilots were reaching Mach 6 and altitudes of over 200,000 feet.

Relative to the time it took to authorize, build, and deliver the X-15, NASA's testing proceeded quickly. Author Dennis Jenkins writes that most of the X-15's original research areas were generally examined by the end of 1961 (though research on these areas continued until 1967). As early as 1961, NASA and the Air Force began casting about for additional "follow-on" research. They eventually approved 46 additional projects ranging from space science experiments (such as ultraviolet stellar photography), to reconnaissance, to component tests (such as a hypersonic ramjet). Follow-on projects eventually made up the majority of all X-15 flights. With additional fuel in drop tanks and an ablative coating, the X-15A-2 reached Mach 6.70, the fastest that the X-15s would ever go. The highest altitude it would reach was 354,200 feet, 67 miles above Earth's surface (almost 10 times higher than commercial jet altitudes). At a total program cost of $300 million (30 times the original estimate), the X-15 program achieved 199 powered flights and inspired over 760 technical documents and reports.[33] The program suffered a number of mishaps, the most tragic being the loss of pilot Michael J. Adams, whose aircraft lost control after entering a high-speed spin. The program ended in 1968, the last flight taking place on 24 October. Proponents of further research, notably Langley's John Becker, sought to modify the aircraft with a delta wing in order to create a hypersonic cruise research vehicle, but this was not to be.

The X-15's technical impact is difficult to measure. Not only was the six-year gestation period an interminably long time in a fast-changing field, but the aircraft itself overshot the performance regime of most other aircraft and missiles. Only three air-breathing aircraft have exceeded Mach 3 (the SR-71, XB-70, and MiG-25), and only a few ventured higher than 50,000 feet. The X-15 did have a valuable impact on the understanding of aerodynamic heating and helped validate wind tunnel methods for predicting such. More spacecraft than aircraft, the X-15 influenced the design and operations of reusable, winged spacecraft such as the Space Shuttle. Arguably, the greatest impact came from the least tangible areas, such as all the tacit learning that went into the design,

33. Jenkins, *Hypersonics Before the Shuttle*, p. 80.

(NASA image ECN-2359)

Figure 3.3. X-24A, M2-F3, and HL-10 lifting bodies on the dry lakebed at the Flight Research Center.

manufacture, and debugging of the aircraft. North American Aviation would later design and build the Mach 3 XB-70 as well as the Space Shuttle. NASA would take its experience with the High Range and create a global tracking and data relay system. Perhaps most important of all, the dual nature of the X-15 helped save the NACA from an Earth-bound end.

Lifting Bodies

Whereas the X-15 began its life as a high-speed aircraft and evolved into a high-altitude, space-touching rocket ship, NASA's lifting-body research sought to create aircraft out of returning spacecraft. This was not a single program, but an evolution of ideas that crossed Center lines, became physical prototypes, and, eventually, earned the support of NASA Headquarters. Though NASA characterized the research as space related, the nature of the work was rooted firmly in the challenges of atmospheric flight. The question confronting scientists and engineers was how to create a flyable spacecraft, something that could ride atop a rocket, orbit Earth, survive reentry, and then return to land on an airstrip. While the researchers, most of them steeped in the study of aerodynamics, found this work to be a logical approach to spaceflight, it was also a way to make aerodynamics and the art of aircraft design relevant in an age of ballistic space capsules. The NASA Centers that most ardently promoted lifting-body research were, not surprisingly, the former NACA Centers that found themselves on the fringes of the nation's space program.

To its champions, the appeal of a flyable spacecraft was self-evident. A lifting body could land at an airstrip, obviating the need for expensive and unpredictable ballistic reentries that ended underneath a parachute. There was also something disconcerting about ballistic reentry as it left heroic astronaut-pilots at the mercy of a flight engineer's trajectory and a flotilla of rescue ships. As Dale Reed, a central figure in lifting-body history, noted, their unofficial motto was, "Don't be rescued from outer space—fly back in style."[34] Lifting bodies would also, like airplanes, be able to fly again. It was taken as an article of faith that operating a spacecraft like an airplane was not only preferable, but also more economical and futuristic. NASA's lifting-body proponents were hardly alone in seeking a reusable, flyable reentry vehicle. The Air Force had begun its own program, the X-20 Dyna-Soar, which sought to create a winged, piloted vehicle that could be launched on a rocket booster and sent on orbital missions.[35] The idea of a winged spacecraft itself had been stock material in rocketry circles for decades, not only in the U.S., but Europe as well. As usual, NASA carried out studies on the X-20 on behalf of the military.[36] The Air Force would eventually cancel Dyna-Soar, just as NASA would eventually cancel further lifting-body research; but these programs, which saw some amount of cross-fertilization and cooperation, would remain important subplots within American civil and military space programs. Of course, flyable reentry became a feature of the Space Shuttle, but it also persisted in on-again, off-again programs, such as the National Aero-Space Plane (NASP or X-30) in the 1980s and X-33 and Venturestar of the 1990s. As recently as 2005, the Air Force continued to lobby

---

34. R. Dale Reed, with Darlene Lister, *Wingless Flight: The Lifting Body Story* (Washington, DC: NASA SP-4220, 1997), p. 17.
35. On Dyna-Soar and the U.S. Air Force, see R. Cargill Hall and Jacob Neufeld, eds., *The U.S. Air Force in Space: 1945 to the 21st Century, Proceedings, Air Force Historical Foundation Symposium, Andrews AFB, Maryland, September 21–22, 1995* (Washington, DC: USAF History and Museums Program, 1998). Also, an entire issue of *Quest: The History of Spaceflight* was devoted to the X-20; see vol. 3, no. 4 (winter 1994).
36. NASA contributions came from both space and aeronautics research. See, for example, Ralph P. Bielat, "Transonic Aerodynamic Characteristics of the Dyna-Soar Glider and Titan III Launch-Vehicle Configuration With Various Fin Arrangements" (Langley Research Center, Hampton, VA: NASA TM-X-809, April 1963).

for a flyable space-plane that could allow the United States to strike globally within a matter of minutes rather than days and weeks.[37]

As bold as the idea of the lifting body was, what distinguished this program from many other equally risky ideas was the degree to which it represented a grassroots initiative rather than top-down planning from Headquarters. In the early years of the research program, there was, in fact, no program. It was merely research championed by a small number of individuals and enabled by discretionary funding from the Centers, primarily the High Speed Flight Station, or, as it was renamed in 1959, the Flight Research Center. Eventually the research did gain attention and funding from Headquarters, but lifting bodies and reusable vehicles were not part of NASA's initial space program, which had set its sights on going to the Moon as soon as possible. Lifting bodies appeared to be a promising area of research for the next generation of space vehicles, and so long as the aeronautical engineers wanted to explore the idea, Headquarters abided.

The intellectual genesis of the lifting body began at Ames with Harvey Allen's work on blunt objects and their ability to shed heat during reentry. Allen, Al Eggers, Clarence Syvertson, George Edwards, and George Kenyon took these ideas further, exploring how to create a blunt-body shape that had useful lift.[38] With lift, such a shape might be able to slow its descent enough to perform an airplane-style landing. Adding wings was one option, but doing so added weight and created surfaces that greatly disturbed airflow and elevated reentry temperatures. Eggers arrived at the idea of using a cone on its side with the top portion sliced off. The bottom portion would shed the heat of reentry while the upper would generate a small but useful amount of lift. The idea worked in the wind tunnel, but it remained a question as to whether such a shape could be controlled and landed.

The work at Ames stimulated engineers at both Langley and, more importantly, the FRC. Dale Reed, an aeronautical engineer who had been at Muroc since 1953, began his own small-scale investigations of lifting bodies in 1962. From paper models to radio-controlled balsa models, Reed grew fascinated by the idea. Drawing together a core of supporters, Reed gained the backing of FRC Director Paul Bikle and Eggers at Ames. Bikle provided funds to the group

37. Tim Weiner, "Air Force Urges Bush to Deploy Space Arms," *International Herald Tribune* (19 May 2005), *http://www.nytimes.com/2005/05/18/world/americas/18iht-space.html* (accessed 17 May 2013).

38. Alfred J. Eggers, Jr., and Thomas J. Wong, "Re-Entry and Recovery of Near-Earth Satellites, With Particular Attention to a Manned Vehicle," NASA Memo 10-2-58A, Ames Research Center, Moffett Field, CA, October 1958.

while Eggers responded with wind tunnel time and advice. In four months, Reed's group hand-built a lifting-body glider large enough to accommodate a single pilot. The most pressing problem they faced was controllability: in such an unconventional aircraft, what should the arrangement of control surfaces be, and how should the pilot use them? These questions would animate the lifting-body program.[39]

In March and April 1963, with an ingenuity characteristic of the FRC, Reed's team began flying the M2-F1 by towing it aloft behind a Pontiac Catalina. In contrast to the FRC's X-15 research, the lifting-body research looked like an overgrown hobby project. Despite its homespun origins, its supporters were both talented and dedicated. After testing the M2-F1 glider in Ames's full-size, 40-by-80-foot tunnel, the group proceeded to higher altitudes using an R4-D tow-plane, the Navy version of the Douglas DC-3. For the next two years, the FRC group, using funds provided by Bikle, tested the M2-F1's flight control characteristics. The last flight took place in August 1966. The M2-F1 gave NASA a baseline for understanding the aerodynamic, control, and piloting characteristics of one type of lifting-body design. Above all else, it convinced many skeptics of the viability of the idea and led to a full-fledged, Headquarters-sponsored program.[40]

Even as Reed and his team plugged away at the M2-F1's research flights, other researchers at Langley and the Air Force were making parallel investigations. At Langley, a group led by Eugene Love had been examining various reentry configurations and had approached the problem differently than Reed's FRC group. Langley's engineers did not limit themselves to lifting-body shapes and examined winged vehicles as well. They arrived at a modified delta shape that was flat on the bottom and curved on the top, nearly the reverse of the FRC design. Rounding out the competition, the Air Force had been experimenting with its own scale lifting body, the SV-5, designed and built by Martin. The Air Force launched models of the SV-5 (as the X-23) atop Atlas boosters to test high-speed reentry. The SV-5/X-23 shape was, in NASA's eyes, halfway between the FRC and Langley designs.[41]

39. Reed, *Wingless Flight*, p. 11.

40. Ibid., p. 65.

41. "[The SV-5 is] a hybrid configuration when compared with the M-2 and HL-10 in that it has somewhat the general appearance and volumetric distribution of the HL-10 but similarity to the M-2 in flight control characteristics." Raymond L. Bisplinghoff and Alexander H. Flax to Associate Administrator; Director of Defense Research and Engineering, DOD; and Co-Chairmen, Aeronautics and Astronautics Coordinating Board, "Interim Report on Coordination of NASA/

With Headquarters' backing and now under the direction of the Office of Advanced Research and Technology, the FRC moved to build two "heavyweight" lifting bodies, the M2-F2 and the HL-10 (the Langley design). Northrop won the contract for the two vehicles in June 1964 and delivered them in 1965 and 1966, respectively, for the bargain price of $1.2 million each. This was partially the result of the FRC's and Northrop's eagerness to reduce bureaucracy and share resources (even when not contractually obligated to do so).[42] The Air Force, meanwhile, also decided to build a piloted, low-speed (relative to the reentry models) version of its SV-5 that was to be named the X-24; so three different candidates headed for testing. In 1965, Bikle was able to fold the Air Force's X-24 into a combined testing program that not only gave NASA access to the Martin design, but also provided launch support and rocket engines for all of the vehicles.

In spite of their differences, all three of the lifting bodies had similar operational designs. As they slowed from their orbital velocities and shed heat, they would need to follow a strict glide path to arrive at the landing strip with just the right amount of kinetic energy. Too much speed and they would overshoot their target. Too little speed and they would be unable to execute a critical last-second flare above the runway. The task was challenging, for not only were these inefficient gliders (with very low lift-over-drag ratios), but their controls were unconventional and potentially unstable. Like the X-15, the lifting bodies launched from an in-flight B-52. At first the tests were unpowered, but, as pilots and engineers gained confidence, the FRC installed XLR11 rocket engines for high-speed tests.

The first flight of the M2-F2 took place in 1966, and testing with the XLR11 rocket engine began the following year. The M2-F2's shape and its early control surface configuration made it prone to Dutch rolls (a combined yawing/rolling motion). Pilot Bruce Peterson crashed the M2-F2 after his efforts to counter a Dutch roll prevented him from executing a proper landing. Northrop repaired and modified the vehicle (renamed the M2-F3), incorporating better lateral stability and additional reaction control jets. It returned to flight in 1970. The aircraft went supersonic under the hand of pilot Bill Dana in August 1971 and it would later achieve a top speed of Mach 1.613. The M2 conducted a total of 43 flights from 1966 to 1972.[43]

USAF Lifting Reentry Program," 20 May 1964, OAST Chron File, "OART Correspondence, May–August 1964," folder 18272, NASA HRC.

42. Hallion, *On the Frontier*, p. 154.

43. Reed, *Wingless Flight*, chap. 5. Individual flight data are available in Reed's appendix.

The HL-10, the Langley candidate, first flew in 1966 and showed significant lateral stability problems caused by flow separation on the vertical stabilizers. After Langley engineers studied the problem and modified the craft, it returned to flight two years later. The aircraft went supersonic in 1969 and reached a top speed the next year of Mach 1.86. The HL-10 completed a total of 37 flights.[44]

Interestingly, it was the Air Force's X-24A that showed the most promise. Flying in 1969, it was the last of the craft to make it into the air. It reached supersonic speeds the following year and a maximum speed of Mach 1.6 in 1971. It completed a total of 28 flights before the Air Force decided to radically modify the vehicle to include a double delta shape. The idea emanated from the Air Force's Flight Dynamics Laboratory and changed the pudgy X-24 into a sleek, triangular craft with an increased lift-to-drag ratio. This second version, the X-24B, flew from 1973 to 1975, reaching speeds of Mach 1.76 and completing landings on a regular concrete runway.[45]

As the lifting-body tests drew to completion, NASA Headquarters refused to authorize any further continuation of the program. Dale Reed, working along similar lines to the Air Force, had been developing a high-lift-to-drag design named the Hyper III. This did not progress beyond an air-launched, remote-control scale model. To some extent, the success of the lifting-body program spelled its end. The FRC team had shown that a variety of unpowered shapes could be safely flown to a runway landing. Where there were problems in configuration and control, these could be hammered out through developmental testing and pilot training. Even though there were engineers who sought to pursue the technology further, there was no significant lifting-body constituency. NASA's decision to pursue a reusable, unpowered winged shuttle (as opposed to a lifting body) after Apollo was both a repudiation and a validation of the lifting-body thesis.[46]

Against the background of NASA's subsequent aeronautical research, the lifting-body program is especially noteworthy because it represented a more traditional, empirical, organic model of research. It was a small, bottom-up project that relied on the enthusiasm of Dale Reed, modest support from Center Director Bikle, and an iterative problem-solving process built on generous testing, including wind tunnel runs, remote-control tests, and piloted tests. The FRC was a safe harbor that allowed Dale Reed to pursue his own interests by fabricating and testing lightweight prototypes. Reed bootstrapped his project by drawing on a wide array of talent at the FRC; Reed's limited

44. Ibid.

45. Ibid.

46. Ibid.

budget did not necessarily constrain his access to expertise and to building an integrated team. The lifting-body program was also very much in the character of the FRC. All of the original Centers had decentralized decision-making and homegrown projects, but the FRC was about flight testing. The approach was hardly dismissive of wind tunnel testing, but for FRC researchers, a problem was not truly solved until it proved itself in the skies above Muroc.

Familiar Speeds and Unfamiliar Roles: The Supersonic Transport

As noted above, NASA did not get, nor did its leadership want, the burden of the nation's supersonic transport program.[47] Going to the Moon was sufficiently taxing, though even if NASA had led the SST program, it is worth considering the novelty of the undertaking. Technically, the "upstream" high-speed research demanded by the SST was right up NASA's alley. But this was a development project; NASA and its predecessor, the NACA, had never attempted to build a viable commercial aircraft. Despite NASA's burgeoning project management skills, an SST project would draw the government much further into the commercial realm than ever before. The SST violated a long-held ideological boundary between public and private investment, becoming an example of both corporate subsidy and the practice of "picking winners." What was the commercial case for an aircraft that private industry was unwilling to underwrite on its own? The SST's ideological risks were as great as its technical hurdles. The government had proved willing to invest in "fundamental" research that could be generally appropriated by private industry: it had funded an aviation infrastructure made of airports and air traffic control; it had established regulatory agencies for the safe and efficient operation of air commerce; but it did not design commercial aircraft.

The SST was supposed to be different. To some, the aircraft was a cornerstone of future national competitiveness and international image. With the British and the French subsidizing the development of their own SST, the U.S. had (according to SST proponents) no choice but to respond. Then there was the aircraft's symbolic importance. For a nation that so identified itself with technological superiority, especially in aviation and ever more so in the era of the Cold War, the SST, at least to a minority, was a matter of national destiny. In 1963, the SST moved forward, not under NASA, but under the FAA. Ultimately, the SST program unraveled as commercial support waned, environmental opposition waxed, and Congress fretted about costs. Congress

47. Hansen has an excellent chapter on the history of the American SST in *The Bird Is on the Wing*, chap. 5.

canceled funding for the SST in 1971.[48] Significantly, the American SST did not become a grand cause for national embarrassment even as the Soviet Union and the Europeans fielded their own SSTs. For a nation that had planted its flag on the Moon, flying passengers at twice the speed of sound in a commercially unviable aircraft triggered little national angst.

The cancellation of the SST obscures the behind-the-scenes commitment to supersonic transports within American aviation. The SST, for a small faction, remains an aircraft that should have been, a tragic missing link in American technological progress. Even in the face of political, economic, and technical setbacks, this idealized technological trajectory has remained durable. The SST vision began, not surprisingly, in military circles. In the 1950s, the U.S. Air Force, for whom strategic bombing was a centerpiece, sought a bomber that could cruise at supersonic speeds at high altitudes, thus evading and penetrating enemy defenses.[49] The first iteration of this was the Convair B-58, a large, four-engine, delta-wing jet designed to deliver nuclear weapons. Though the B-58 suffered from high fuel consumption, limited payload, and only supersonic dash capability, as opposed to a sustained cruise, the Air Force eventually fielded two B-58 bomber wings. The Air Force, predictably, sought a more capable replacement and began the development of the B-70, a Mach 3 design from North American Aviation. In 1959, Eisenhower, unconvinced of the worth and logic of the B-70, canceled the program. Although the Air Force maintained the development of the aircraft as an experimental program, this turn of events was ultimately the start of a long, slow death for large, supersonic aircraft. U.S. industrial policy, insofar as the country had a policy, existed to support military technological development. A commercial supersonic transport might have been a serendipitous byproduct of the military bomber program, just as earlier military jets forged a path for subsonic commercial transports. Lacking a bomber program, supersonic advocates had to figure out how to support a high-risk development program outside the military.[50]

Even though NASA did not take up the SST project in the 1960s, it was very much at the center of the supersonic movement. As noted in chapter 2, Langley researchers had been instrumental in providing design guidelines for supersonic aircraft. The B-58 was, in some respects, a scaled-up version of the Convair F-102, the aircraft that sported Richard Whitcomb's Coke-bottle

48. Horwitch, *Clipped Wings*, chap. 19.

49. The Air Force also studied the idea of supersonic transports in addition to bombers. Conway, *High-Speed Dreams*, p. 40.

50. Conway, *High-Speed Dreams*, chap. 1.

area-rule fuselage. North American's design for the B-70 borrowed heavily from research performed by Al Eggers at Ames, notably the idea that a supersonic aircraft could ride atop its own shock wave. Years before the start of the American SST program, in 1956, John Stack and his colleagues at Langley had begun looking at commercial SSTs. By 1960, they were vocal advocates of a NASA-led SST program and formed a Supersonic Transport Research Committee. Far from being dispassionate theorists and experimentalists responding passively to military and commercial requests, researchers like Stack strongly promoted specific technological trajectories.[51]

What Stack and his peers lacked was high-level political support, but here leaders in the newly formed FAA came to the rescue. Established the same year as NASA, the FAA began examining SSTs in 1959 and quickly developed ties to Langley's cohort of SST champions. The FAA's first administrator, Elwood Quesada, attempted, unsuccessfully, to sell the SST to a dubious Eisenhower. President Kennedy's FAA administrator, Najeeb Halaby, had more luck. In 1962, the FAA and NASA contracted to Boeing and Lockheed for exploratory studies on possible SST configurations, that is, potential wing-fuselage-engine arrangements and projected performance specifications. In 1963, Kennedy gave Halaby a very limited SST program that maintained funding for exploratory studies at NASA and the aircraft manufacturers. There can be little doubt that the Anglo-French announcement to pursue an SST in November 1962 prodded the American response. In retrospect, it is evident that the American response was not at all symmetrical. Where Europe sought to keep its industry and prestige alive in the face of very real American market and technological domination of commercial jet transports, the Americans were merely hedging their bets, keeping the SST in play without fully committing to the technology or its commercialization.[52]

Through the 1960s, the FAA-led program examined a number of approaches. How fast and high should the aircraft fly? What markets should it serve (i.e., what distance and how many passengers)? What configuration represented the optimum balance of risk, economic return, and manufacturing cost? How would sonic booms curtail either the design or operation of the aircraft? In the end, many of the technical debates became moot. Studies by the Air Force and NASA, appropriately named Operation Little Boom and Operation Bongo,

51. Stack's group was not alone in its enthusiasm for SSTs. Indeed, the British and the French had established research programs in the previous two years. Conway, *High-Speed Dreams*, pp. 37–38, 54. Hansen, *The Bird Is on the Wing*, p. 152.
52. Hansen, pp. 152–154. See also Kenneth Owen, *Concorde and the Americans: International Politics of the Supersonic Transport* (Washington, DC: Smithsonian, 1997).

suggested that the public was not at all enamored with regular supersonic booms.[53] Not only did this limit the technology to transoceanic flights, but SSTs would have to fly the beginning and ending portions of their trips at inefficient subsonic speeds. Even more crippling were the economic projections. Despite initial enthusiasm and public proclamations of an SST future, airlines were not willing to commit themselves. Likewise, as the manufacturers delved deeper into the technical and economic details, they became increasingly unwilling to shoulder the aircraft's risk. This was critical because one of the conditions of the Kennedy, and later Johnson, administration's lukewarm support for SST was that industry share in the vehicle's development costs. As historian Erik Conway notes, the environmental opposition that the SST faced in the early 1970s merely drove one more nail into the coffin.[54]

The technical challenges of the SST resulted in prolonged debates and not a small amount of testing. There were long-term fruits to this work even though the intended product never materialized. Of prime importance was the question of how to design an aircraft that operated efficiently across a number of very different flight regimes. Takeoff and landing took place at around 200 miles an hour; climb-out and descent to landing would need to be at subsonic speeds; and cruise would be at supersonic speeds at an altitude of at least 40,000 feet. An aircraft that performed well in one regime often performed poorly in the others. NASA's engineers had many different ideas about the optimal configuration, but in 1963 they asked industry to evaluate four designs. Of these, three originated from Langley and one from Ames. Two of the designs were fixed-wing, and two were variable-geometry, also known as swing-wings. The Germans had toyed with the idea of swing-wings during World War II, and Bell had built an experimental vehicle (the X-5) that the NACA and the Air Force tested in the 1950s. A swing-wing could be angled forward (low sweep) for low speeds and angled backward (high sweep) for supersonic speeds. Unfortunately for the four NASA designs, industry found them wanting, at least against the specified performance criteria.[55]

The FAA, for its part, was already pushing ahead with its own plans, publishing in 1963 its request for SST proposals; Boeing, Lockheed, and North American all submitted designs. Boeing adopted a variable-geometry configuration; Lockheed chose a double delta, a tailless design incorporating a fixed

---

53. Hansen, *The Bird Is on the Wing*, p. 155; Conway, *High-Speed Dreams*, pp. 60–61.

54. Both Conway, *High-Speed Dreams*, and Horwitch, *Clipped Wings*, cover environmental opposition to the SST.

55. Hansen, *The Bird Is on the Wing*, chap. 5; Conway, *High-Speed Dreams*.

(NASA image A-37700)

Figure 3.4. A double-delta-winged SST model installed in the Ames 40-by-80-foot wind tunnel in 1966.

triangular wing; and North American chose a delta wing with a canard.[56] NASA and the FAA evaluated the designs, and in 1966 the FAA named Boeing the winner. Though NASA's researchers preferred their own configurations to those of industry, there was, to Boeing's benefit, strong support from Langley for a variable-geometry solution.[57] Boeing won the competition, but as the company refined the design and performed new tests and calculations, it found that the aircraft was becoming too heavy to meet the FAA's specifications for range and payload. As the design evolved, problems multiplied. Boeing eventually gave

56. The double delta wing incorporated two fixed sweep angles: a narrow high-sweep wing combined with a wider low-sweep wing. Imagine a tall, thin triangle superimposed over a short, squat triangle.

57. Hansen, *The Bird Is on the Wing*, pp. 157–161. Langley's enthusiasm for variable geometry originated in work it had done in the late 1950s for testing and refining British swing-wing designs. Langley's team adopted the British idea of placing the pivot outside the fuselage and refined it by housing the pivot in an inboard wing. See Conway, *High-Speed Dreams*, pp. 53–54; and Edward C. Polhamus and Thomas A. Toll, "Research Related to Variable Sweep Aircraft Development" (Langley Research Center, Hampton, VA: NASA TM-83121, May 1981).

up on the swing-wing design in favor of a slender, triangular wing, not unlike Lockheed's losing entry.[58]

Configuration was one of many technical question marks hanging over the American SST. The propulsion system, for example, was equally experimental, as were solutions to the problem of aerodynamic friction. Though designers ultimately chose to build out of traditional aluminum alloys, operational temperature extremes meant that engineers had to redesign many systems created for subsonic jets. What boded ill for the SST, above all else, was the inability of Boeing to meet the FAA's performance and cost specifications. Fortunately for the company, Congress voted to cancel the program before there was a chance to see whether the latest design was successful.[59] As is discussed in the next chapter, the SST ran headlong into a burgeoning environmental movement, a popular and political shift that impacted not only supersonic transports, but NASA research as well.[60] For NASA's SST researchers who understood the shortcomings of Boeing's SST, the cancellation of the program was a setback not just because the aircraft never flew, but because the budget for fundamental supersonic research collapsed as well.[61]

Subsonic Aerodynamics

One of the conundrums of NASA's research is that although big, expensive projects like the X-15 grabbed headlines and became the public face of NASA's aeronautical work, it was the smaller, less dramatic, less expensive research that was both more representative of the Centers' work and, arguably, more critical to the growth of aeronautical knowledge. These relatively small technical reports had, and continue to have, a difficult time competing with risky, vehicle-centered projects. While much 1960s research continued to "push the envelope" of what was aerodynamically possible, such as hypersonics, work also continued on more pedestrian subsonic speeds. The best example of this work was the refinement of subsonic wing design at Langley. Most subsonic jet aircraft built today now employ an intellectual descendant of Langley's wing designs.

58. For an in-depth examination of the evolution of the SST's configuration, see Conway, *High-Speed Dreams*, chap. 3.

59. James R. Hansen, "What Went Wrong?: Some New Insights into the Cancellation of the American SST Program," in From Airships to Airbus, ed. Leary, pp. 168–189. Hansen notes that Boeing was also bringing its first wide-body jet to market, the 747 jumbo jet, and that any SST sales were potentially lost 747 sales.

60. Conway, *High-Speed Dreams*, chap. 4; Horwitch, *Clipped Wings*, chap. 19.

61. Hansen, "What Went Wrong?" pp. 175–180.

Although subsonic aerodynamics was perhaps the best understood of all flight regimes, the area retained special challenges and opportunities. The commercial jet industry, of course, eagerly sought improved designs. Industry also demanded that innovations balance improved flight efficiency with attendant manufacturing and operating costs. Unlike X-planes and many military aircraft, performance increases were tightly constrained by competitive airframe and airline markets. By their nature, new designs for commercial aircraft had to exhibit a level of safety and reliability unnecessary for advanced military aircraft. Finally, because of the low cost of manufacturing tubular fuselage sections of constant width (relative to highly optimized variable-diameter sections), aerodynamic advancements had to be found elsewhere, principally the wings.

There were important breakthroughs in the development of advanced subsonic wings, but there was no single solution because the problem took on many forms. It is not terribly difficult to design an airfoil (a two-dimensional cross section) that is perfectly optimized for a given altitude and speed. But aircraft must take off and land, cruise at different altitudes, maneuver, burn fuel as they fly (thus reducing their mass), and be designed for a variety of tasks. Neither theory nor one-off designs would satisfy industry. Langley researchers needed to provide a range of proven design options, as well as the tools necessary to create three-dimensional shapes. Truly, this kind of work was in the great tradition of the NACA's earlier contributions to wing design, especially the NACA airfoils that allowed manufacturers to pick and choose designs as though from a catalog. What resulted, in the end, was a combination of evolved techniques developed by scores of researchers.

The cornerstone of subsonic wing development was what became known as the "supercritical wing," though this too represented both a collection of design choices and methods for optimizing performance. Richard T. Whitcomb, who had established his reputation with his area-rule research in the 1950s, led a team of researchers in examining high-speed subsonic airfoils. Whitcomb's own contribution was both his intuitive style of wind tunnel experimentation and his credibility. Amidst a number of competing research programs as well as inherent conservatism from industry, Whitcomb's advocacy helped mobilize NASA engineers and attract commercial interest. At the time Whitcomb undertook this strand of research, the state of the art in subsonic wing design was an airfoil from the British National Physical Laboratory (NPL). Compared to previous airfoils, the British one had a flatter camber, which delayed and reduced the shock wave that formed on the top of the wing. Such shock waves added drag, known as wavedrag. NPL-derived airfoils would appear in jets designed in the late 1960s.

Whitcomb, who had grown disenchanted with the direction that the SST was taking, struck out to find a new research program. His inspiration came after examining the performance of a vertical-takeoff wing from Ling-Temco-Vought. The engines forced air out of a slot on the upper surface of the wing with an effect similar to that of the NPL airfoil. From this, Whitcomb formed a research group to examine the problem of wavedrag on subsonic wings. The project lacked the prestige of supersonic and hypersonic work, but Whitcomb saw it as potentially more fruitful, and his boss, Laurence K. Loftin, gave him carte blanche. Whitcomb's starting point was the realization that wavedrag propagated forward from the trailing edge. The NPL and Ling-Temco-Vought wings reduced the wavedrag because they delayed airflow separation. Whitcomb's team began modifying airfoils, testing them in the 8-foot transonic, and analyzing schlieren images afterward to visualize the airflow.[62] Because Whitcomb did not begin with a theory or mathematical model, his methods have been called cut-and-try. Whitcomb bridled at the description, for each change in the airfoil was guided by a deep understanding of what was occurring and how it might be solved. The first effective design incorporated a trailing-edge slot; air from underneath the wing would flow to the top and help maintain a smooth stream. The team dispensed with the slot because of its structural complexity and settled upon an airfoil with a blunt leading edge, a flattened upper camber, and an exaggerated camber underneath the trailing edge to compensate for the loss of camber on the top.[63]

The appellation "supercritical wing" is deceptive. It suggests a single, revolutionary idea when, in fact, it was a combination of incremental changes, none of which were entirely new to the field of aerodynamics. Further, the supercritical wing was not born of theory, but inspired and perfected through hands-on wind tunnel testing. What the name reflects, in part, was a second part of Langley's research agenda. If the first part of any research was to mobilize support for an idea within the Lab, the second was to mobilize a winning idea outside the Lab. Here, Loftin not only suggested to Whitcomb that he give the airfoil a name but later called for the development of theoretical approaches to the supercritical airfoil that would allow designers to create such wings without Whitcomb's intuitive wind tunnel methods. At the same time, Loftin pushed for flight tests to validate the airfoils. For the first test, Langley

62. Schlieren imagery makes visible the pressure gradients in an airstream.

63. Richard T. Whitcomb, interview by Robert G. Ferguson, telephone recording, 30 November 2005, NASA HRC. See also John Becker, *The High Speed Frontier: Case Histories of Four NACA Programs, 1920–1950* (Washington, DC: NASA SP-445, 1980), chap. 2; Richard T. Whitcomb and Larry R. Clark, "An Airfoil Shape for Efficient Flight at Supercritical Mach Numbers" (Langley Research Center, Hampton, VA: NASA TM-X-1109, July 1965); and Chambers, *Concept to Reality*, pp. 7–19.

(NASA image EC73-3468)

Figure 3.5. Whitcomb's supercritical wing flown on a modified F-8 in 1973.

and North American Rockwell modified a Navy trainer, the T-2. Normally, the T-2 had a 12 percent–thick wing (i.e., the thickness was 12 percent of the mean chord, the average distance between the leading and trailing edges). With a supercritical airfoil, the T-2 could have a 17 percent thickness with the same performance as the standard wing. The next set of tests, initiated in 1968, called for a joint FRC-Langley program using a Navy F-8 aircraft with a new, supercritical wing. From 1971 to 1973, the modified F-8 flew 86 tests, proving that the technology could allow for higher subsonic speeds and thus greater aerodynamic efficiency.[64]

The first applications of the technology appeared in military aircraft such as the General Dynamics F-111. For commercial aircraft, manufacturers chose to use the airfoil to reduce wing sweep, which itself is a design technique used to mitigate the rise of supersonic drag, creating lighter, more efficient wings. The first commercial application was in the Cessna Citation III business jet, introduced in 1982, and the first substantial use for large aircraft was in the Boeing 777, which first flew in 1994. More recently, the Airbus A380, the largest commercial passenger jet yet, uses supercritical technology to decrease wing sweep to 33.5 degrees. By contrast, the Boeing 747's wing sweep, originally designed in the 1960s, is 37.5 degrees.[65]

---

64. According to Whitcomb, these later tests added little to his own understanding of wing design. He was largely finished with the project by the time the F-8 flew. See Whitcomb telephone interview.

65. For applications of supercritical airfoil technology, see Chambers, *Concept to Reality*.

The Rogallo Wing

The Rogallo wing is a fascinating story that almost would be a side note to NASA's history were it not for the remarkable path the technology took from a home project, to potential space race technology, to becoming the basis for an entire genre of personal aircraft. Francis Rogallo joined the Langley Laboratory in 1936 after earning a degree in aeronautical engineering from Stanford University. In the mid-1940s, Rogallo began to develop the idea of a flexible wing, something that could be made from cloth. According to his own account, the NACA was uninterested in the idea, so he pursued the idea in his spare time and with the help of his family. One of his critical innovations was that the shape of the wing could be maintained by a combination of the load and aerodynamic forces. Rogallo and his wife applied for a patent in 1948 and through the 1950s sought interest from industry. All they managed was to sell a few thousand toy kites based on the design.[66]

A technology in search of an application, the flexible wing found a potential use with the launch of Sputnik. Perhaps a lightweight flexible wing could be used to bring a returning spacecraft in for a landing, unfolding atop the capsule as it dropped through the atmosphere. Interest grew at Langley, and in 1958 the Lab established a Flexible-Wing Section. Rogallo's team studied a number of different unstiffened and stiffened cloth wings. Most of the studies involved wind tunnel analysis, but the Flight Research Center tested a few of the designs. Using a conical wing stiffened with metal poles, the Paraglider Research Vehicle (Parasev) glider made hundreds of piloted flights from 1962 to 1964, towed aloft either from the ground or by another aircraft. In 1965, NASA tested another conical wing, this one stiffened by inflatable tubes and carrying a Gemini-type capsule. The military also took an interest in the early 1960s and experimented with a powered vehicle, the XV-8 for Flying Jeep (or Fleep), but NASA had by 1964 settled on recovering capsules with parachutes and ocean splashdowns, and funding for continued flexible-wing research soon came to a close.[67] The Rogallo wing, however, did not fade into obscurity like so many other ingenious designs. Popular accounts inspired outsiders to design

66. Francis M. Rogallo, "NASA Research on Flexible Wings" (International Congress of Subsonic Aeronautics, New York, NY, 3–6 April 1967); Francis M. Rogallo et al., "Preliminary Investigations of a Paraglider" (Langley Research Center, Langley Field, VA: NASA TN-D-443, August 1960); Robert Zimmerman, "How To Fly Without a Plane," *Invention & Technology* 13, no. 4 (spring 1988); Hansen, *Spaceflight Revolution*, pp. 380–387.

67. Hansen, *Spaceflight Revolution*, pp. 385–387.

their own flexible wings using the basic outlines of the Rogallo wing. These were the first modern hang gliders.[68]

NASA's First Decade

Contrary to the jeremiads that predicted the death of aeronautics at the hands of the space program, aeronautics research persisted through the 1960s. The space program did steal the limelight and garnered the bulk of the Agency's funding, but it also left aeronautics largely to its own devices, a beneficiary of benign neglect. In such an environment, Center management and laboratory practice of aeronautics at Langley, Ames, and the FRC remained much as they had in the latter years of the NACA. The Lewis Laboratory, as noted, devoted itself largely to the space program after 1958. Even as Headquarters mildly attempted to increase oversight throughout the 1960s, the Centers retained the ability to initiate experiments and allocate funding. The flip side to such freedom, however, was that aeronautics research was uncoordinated and lacking top-level leadership. The one project that might have served as a rallying cry, the SST, went to the FAA and became mired in politics and impossible economics. Aeronautics in the 1960s was very much propelled by the momentum of an earlier era. By the late 1960s, however, vestiges of the NACA way were truly coming to an end. Two developments brought this about: NASA Headquarters chose to replace the heads of Langley, Ames, and the FRC with outsiders; and, with Americans on the Moon and the Apollo program drawing to a close, NASA scrutinized, for the first time since 1958, its aeronautics programs. In doing so, it began exporting to aeronautics the managerial centralization and research programming that had been characteristic of the operational space program.

68. Stéphane Malbos and Noel Whittall, eds., *And the World Could Fly: The Birth and Growth of Hang Gliding and Paragliding* (Lausanne, Switzerland: Fédération Aéronautique Internationale, 2005). NASA's 1976 issue of *Spinoff* noted that a California hang glider company was using a Rogallo wing design.

Chapter 4:

Renovation and Revolution

At the beginning of the 1970s, NASA should have been riding high on the success of the Apollo program. In a decade, it had learned to take humans to the Moon and back and in so doing had become the paradigm of modern technological capability. For the United States, it was a fantastic Cold War triumph, but it was also a tremendously ironic accomplishment against a background of social discontent over the fruits of science, technology, and industry. Pollution, ecological damage, energy shortages, and the arms race framed a public discussion of technological progress amplified by an unpopular war in Vietnam, liberal youth expression, and unsettling economic competition from Asia. NASA was hardly complicit in these problems and societal tensions, but as one of the most visible agents of federally funded research, it risked being out of step with the public will. With the Stars and Stripes planted on the Moon, continued lunar missions merely threw salt on the wounds of the country's more Earth-bound concerns.

Without a space race, NASA began casting about for a new purpose and a renewed sense of relevance.[1] NASA's leaders considered transforming NASA into a "technology agency" that would assume "government-wide responsibility for the application of technology to national needs."[2] There was strong support from the White House for NASA to become more business-minded, to evaluate its technological resources from a commercial perspective.[3] This was a

---

1. The end of the space race spelled, in part, a reduced overall budget at the Agency. See James C. Fletcher to Roy L. Ash, 9 July 1973, Fletcher Correspondence, 1972, folder 4248, NASA HRC. Morale at the Agency was also low (see Center Director's Meeting Minutes [handwritten], 11 September 1972, Fletcher Correspondence, 1972, folder 4248, NASA HRC).
2. George M. Low to James C. Fletcher, memorandum for the Administrator, "NASA as a Technology Agency," 25 May 1971, James C. Fletcher Correspondence 1971, folder 4247, NASA HRC.
3. James C. Fletcher to George M. Low, "Luncheon Conversation with Pete Peterson," 20 July 1971, James C. Fletcher Correspondence 1971, folder 4247, NASA HRC. Peter George Peterson was Assistant to the President for International Economic Affairs; in 1972, he was named the Secretary of Commerce.

long way from the fractious debates of the NACA days about undue influence from private industry, and it shifted NASA's charge away from doing basic or fundamental research.

Arguably, for the first time since the creation of NASA, the Agency began to pay attention to its aeronautics program. Aeronautics was practical and economically applicable in a way that astronauts and daring space exploits were not (or at least not obviously so to the public). Aeronautical research could also address public concerns about the environment and energy efficiency. Higher and faster was, at least as a headline, replaced with quieter, cleaner, and more economical. Congress and the White House encouraged Headquarters' newfound interest in aviation. Aeronautics always had friends on the Hill, and the post-Apollo era provided an opportunity for supporters to charge that NASA had neglected aeronautics during the space race. The initial outlines for NASA's civil aeronautics research emerged from a joint Department of Transportation–NASA study, the Civil Aviation Research and Development Policy Study (CARD). The CARD study provided NASA's leadership, the White House, and Congress with an informed basis for supporting a modest expansion in the aeronautics budget.[4]

At Langley, Ames, and the Flight Research Center, the shift toward the pragmatic was less evident than it was in Washington, DC. While Lewis went through a post-Apollo upheaval, the other Centers retained their own technological momentum.[5] Quieter and cleaner did not spell the end of supersonic research or vertical takeoff and landing. This momentum derived from embedded skills and experimental equipment, as well as a certain amount of technological boosterism. Just as the human space program persisted in the face of cyclical public support, so too did a number of long-running themes in aeronautics. The success of these programs relied on researchers' ability to cobble together sufficient political support, whether this was seed money from a Center Director or interest from the Pentagon and industry. The source of funding, project name, justification, and eventual application were, arguably, secondary to keeping the research going.

For all of Headquarters' worries about public support and the contribution of NASA's research to the public good (NASA began publishing the self-promotional report, *Spinoff*, in 1976), researchers did not need to be told how to be innovative. Aeronautics' leap into computers is illustrative of the

4. James C. Fletcher to George P. Shultz, 30 September 1971, James C. Fletcher Correspondence 1971, folder 4247, NASA HRC. George Shultz was the Director of the Office of Management and Budget at the time.
5. Dawson, *Engines and Innovation*, p. 201.

Centers' initiative.[6] The Ames Research Center, building on work initiated in the 1960s, made a bold move into computational fluid dynamics (CFD) as a way of modeling airflow digitally. Given the complexity of fluid dynamics, CFD was not an obvious or simple application of computers; CFD had more than its share of doubting Thomases. Ames researchers gambled and, after working through years of teething trouble, began to show that CFD could displace some wind tunnel work. Ames was not the only laboratory pursuing CFD; indeed, Langley competed in the area as well, and all of the Centers sought to use digital computing to revolutionize research and flight. For Ames, however, CFD offered an attempt to stake out a new field, to occupy a new niche. Competition among the Centers, not top-down programming, encouraged risk taking. Yet even though the Centers had this organic economy, so to speak, with built-in incentives for risk taking, the trend was toward taking these decisions out of the hands of the labs.

One of aeronautics' major public projects of the 1970s was the Aircraft Energy Efficiency (ACEE) program. Viewed from the researcher's perspective, the ACEE program was merely the repackaging of existing research strands under the banner of efficiency. Researchers were hardly calculating opportunists. They really did have a supply of genuinely useful ideas to combat the rise in fuel prices, but viewed from a managerial perspective, ACEE appeared to be a top-down call for answers to a pressing national concern, namely the supply of petroleum. In spite of the continuity in research, the ACEE program gave license to increased program control, not only from NASA Headquarters, but from Congress, too. Aeronautics was no longer operating underneath the radar, and it was no longer immune to the kind of political programming that had been part of the space program. The ACEE program also departed from what had been the NACA's reluctance to fund (or appear to be funding) corporate R&D. While the NACA and NASA had a long history of contributing to the development stage of military projects, they took a more hands-off approach to assisting manufacturers with commercial products. The ACEE changed this and specifically funded (usually to a level of 90 percent) the design, manufacture, and testing of efficiency-related products by the manufacturers. Although there were good reasons for bringing the manufacturers on board (e.g., ease of

---

6. Nearly two decades later, a Congressional Budget Office report, commenting on the idea of spinoff at NASA in general, noted, "The spin-off occupies a central place in the mythology of NASA's relation to the private economy and, accordingly, in the argument that secondary economic benefits might justify spending for NASA." U.S. Congress, Congressional Budget Office, "Reinventing NASA," March 1994, p. 4.

technology transfer), that this methodology came at the request of Congress raised questions of patronage and corporate subsidy.

NASA's renewed interest in aeronautics in the 1970s fundamentally did not alter aeronautics' funding or its relationship to the space program. Where NASA's space budget declined significantly as development of the Apollo hardware matured, aeronautics funding continued with modest increases (in real dollars); in constant dollars, funding tended to be flat in the early 1970s. There was, however, questioning from the Office of Management and Budget (OMB) about the federal government's proper role in aeronautics research. OMB encouraged NASA to share more costs with industry, transfer functions to industry, and require that the armed services fully reimburse NASA for the work performed on military projects.[7]

As in the 1960s, aeronautics and space continued to have mutual interests. In the post-Apollo era, NASA moved to define a new space transportation system, one that evolved into the Space Shuttle.[8] Here, earlier work in lifting bodies contributed to the Shuttle's reentry design. For the 1970s, the second phase of the digital fly-by-wire program would adopt the Shuttle's computer system and, in turn, make important contributions to the design and testing of the Shuttle's mission-critical software. As aeronautics' work on computational fluid dynamics matured, it became as important to the simulation of spacecraft reentry as it did to conventional flight. The Shuttle, by virtue of being a hypersonic glider, gave hope to some that the spacecraft would serve as a platform for aeronautical research. At a top-level meeting of OAST in 1972, managers wrote, "The [S]huttle carries the potential of being a stepping stone beyond the X-15 research airplane; a stepping stone to open new regimes of aerospace effectiveness supplementary to other hypersonic research airplanes."[9]

The political-economic context for the Agency in the 1970s was dominated by the oil embargo, energy shortages, and growing environmental awareness. Another economic factor would have a deep impact on the nation's airlines and, indirectly, NASA: in 1978, the Civil Aeronautics Board deregulated the airlines. Up to that point, 10 airlines controlled some 90 percent of the nation's market with routes strictly apportioned. On thinly traveled routes, the government guaranteed

7. George M. Low to James C. Fletcher, "Discussions with Glenn Schleede," 14 November 1974, James C. Fletcher Correspondence 1974, folder 4250, NASA HRC.
8. On the early design of the Space Shuttle, see T. A. Heppenheimer, *The Space Shuttle Decision: NASA's Search for a Reusable Space Vehicle* (Washington, DC: NASA SP-4221, 1999).
9. Deputy Associate Administrator, Management Office of Aeronautics and Space Technology, to OAST Management Council et al., "OAST Management Council Meeting and Spring Planning Conference held May 24–25 at Goddard Space Flight Center," 16 June 1972, OAST Chron Files, "Gen/RD–T Chron File 1972," folder 18274, NASA HRC.

airlines a specific profit margin. The era of deregulation opened the door to new startups, especially low-fare airlines. Route structures changed dramatically as most airlines migrated to hub-and-spoke networks and reduced service to remote locales. The arrangement increased efficiencies but at the cost of fewer direct flights (and in some cases, no flights). Airlines shifted to smaller jet aircraft and increased flight frequencies in order to better compete. These moves exacerbated airspace congestion at the nation's busiest airports. Deregulation greatly aided the rise of the regional turboprop and jet market segment, dominated by two foreign manufacturers, Embraer of Brazil and de Havilland/Bombardier of Canada.[10]

Reverberations from the Vietnam War dominated the military context for NASA. Rotorcraft were now central to the operation of ground forces, and the Army took the lead in fostering research on this front. In contrast to the U.S. Air Force's post-WWII gambit for its own R&D capabilities, the Army sought more modest capabilities and did so in close cooperation with NASA. In terms of fixed-wing aircraft, a battle was playing out in the Pentagon over what constituted the best fighter design. More traditional elements preferred large, twin-engine air-superiority aircraft, while a renegade group advocated small, more highly maneuverable aircraft that traded size and complexity for agility.[11]

Coordination and Research Directions

Organizationally, the 1970s were consistent with the long-term trends of greater centralization and managerial control over research programs. At the Headquarters level, this is well represented by a somewhat wishful diagram penned for a 1971 management council meeting (figure 4.1). It shows the Office of Advanced Research and Technology as a railroad locomotive and the four OART Centers (Langley, Ames, Lewis, and the FRC) as freight cars. In the "before" half, all of the cars are traveling in the same direction, seemingly flying through the air, but none hitched to another. In the "after" half, the OART locomotive pulls the Centers, all dutifully aligned, coupled, and on the same track. Such coordination was, at least for the 1970s, more hope than reality, but it does capture the desire to centralize

---

10. Louis Uchitelle, "Off Course," *New York Times Magazine* (1 September 1991): 12–16; Elizabeth E. Bailey, David R. Graham, and Daniel P. Kaplan, *Deregulating the Airlines* (Cambridge, MA: MIT Press, 1985); Steven Morrison and Clifford Winston, *The Economic Effects of Airline Deregulation* (Washington, DC: Brookings Institution, 1986). An interesting question one might ask is whether regulatory structures of the 1950s and 1960s indirectly led American airframe manufacturers to abandon small regional aircraft.
11. Richard Hallion, "A Troubling Past: Air Force Fighter Acquisition Since 1945," *Airpower Journal* (winter 1990); Robert Coram, *Boyd: The Fighter Pilot Who Changed the Art of War* (New York: Little Brown and Company, 2002).

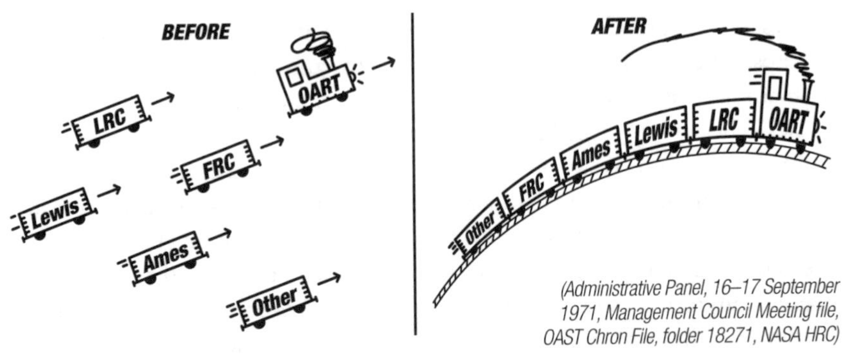

Figure 4.1. This diagram, based on a drawing from a management council meeting, 16–17 September 1971, depicts the Office of Advanced Research and Technology as a locomotive pulling the Centers.

research underneath OART. Management council meetings, begun in the early '70s, were obviously one tool to this end. Of greater consequence, at least from a top-down perspective, was the increase in directed research with priorities set in Washington, DC. The environmental and energy-efficiency programs were emblematic of this trend. As noted in the previous chapter, the 1970s brought a new generation of leaders to the Centers. In the case of both Langley and Ames, these Center Directors were the first to come from outside the NACA/NASA ranks. Such new blood brought new energy and initiative, and in both cases the Directors made lasting and contentious changes.

The degree to which NASA sought to apply itself to contemporary social issues is nicely exemplified in a 1971 letter from OAST Associate Administrator Roy Jackson to the Department of Education, offering NASA's assistance on "the solution of certain public sector problems." He noted their research at one California high school that sought "to determine if school disorders could be reduced," namely through an alarm system for teachers and "an automated student attendance counting system."[12] In a letter to the Associate Administrator for Space Science Applications, Jackson noted, "There has been considerable public interest generated during the past year concerning the extent to which our Aeronautics and Space resources are being used to help federal, state and municipal agencies."[13] At least at Headquarters, building bridges to other agencies was the order of the day. In conjunction with the National Science Foundation,

12. Roy P. Jackson, AAOART, to Dr. Sidney P. Marland, Jr., Commissioner of Education, U.S. Office of Education, 15 October 1971, OAST Chron File, "RF Reading File 11/1/71 to 12/31/71," folder 18274, NASA HRC.
13. Roy P. Jackson to AA for Manned Space Flight, AA for Space Science Applications, 4 August 1971, OAST Chron File, "RF Reading File 11/1/71 to 12/31/71," folder 18274, NASA HRC.

NASA agreed to evaluate "the extent to which aerospace-derived technology and capability can assist cities in meeting rising technology-related problems."[14]

In 1972, the Office of Advanced Research and Technology changed its name to the Office of Aeronautics and Space Technology, a switch that was more rhetorical than substantive. It reflected, however, the Agency's intention to cast itself as a problem solver rather than an ivory tower laboratory. Managers at Headquarters were certainly under pressure to see that NASA delivered goods to a wider audience. It could no longer be narrowly concerned with the competitiveness of American aeronautics and placing humans on the Moon. By 1974, the rhetorical shift in OAST's mission was finally enunciated:

> The NASA Aeronautics Research and Technology programs are directed at serving national needs by focusing on the objectives of developing technology to (1) reduce energy requirements and improve the performance and economy of aircraft, (2) reduce the undesirable environmental effects of aircraft such as noise and pollution, (3) improve safety and terminal area operations, (4) advance short-haul, short takeoff and landing, and vertical takeoff and landing system concepts, and (5) provide aeronautical technology support to the military.[15]

In spite of all the shuffling going on at Headquarters, the budget mix for aeronautics remained recognizable (see table 4.1).

Table 4.1. Aeronautical Manpower Distribution, FY 1972

Division	Number	Percent
Aero Propulsion	1,344	38.5
Aero Vehicles	1,151	32.7
Materials and Structures	344	9.8
Aeronautical Operating Systems	243	6.9
Advanced Technology Transport	192	5.2
Guidance, Control, and Information Systems	131	3.7
Aeronautical Life Sciences	55	1.7
STOL	50	1.5

Source: "Program Manpower Assessment Panel Report" (presented at the Management Council Meeting, 16–18 September 1971), Management Council Meeting file, OAST Chron File, folder 18271, NASA HRC.

14. R. D. Ginter to Assist. AA for Advanced Research and Technology, "RF Reading File 11/1/71 to 12/31/71," OAST Chron File, folder 18274, NASA HRC.
15. NASA, *Aeronautics and Space Report of the President, 1974 Activities* (Washington, DC, 1976), p. 31.

Prior to the oil embargo, NASA's emphasis on "environmental" problems focused primarily on noise reduction. The SST, of course, figured prominently in NASA's 1960s-era noise reduction work, but the growth of commercial jet aviation in general (and the rise in airport traffic) prompted growing complaints from residents near airports.[16] The oil embargo served as a catalyst for a larger rethinking of NASA's relationship to the environment. By late 1973, at a Center Directors' meeting with Administrator Fletcher, the group decided to establish a small group "to review and understand the results of existing studies of the total energy problem, and to identify those areas of energy related technology where NASA might make a significant contribution." Thinking specifically about aeronautics, they decided to focus future advanced technology efforts on minimizing fuel consumption and to study what could be done with existing aircraft. More generally, the group considered the following "Super Problems":

a. Greater support for the Defense Department
b. Nuclear waste disposal
c. Communications
d. Manufacturing in space
e. Global food supply
f. Depletion of our natural resources
g. Global environmental surveys and earth resources surveys
h. A plan to bring the world into equilibrium by the year 2050
i. A "whole planet approach" to solving the earth's problems[17]

Another way to make NASA relevant was to strengthen ties to industry. The NACA had long partnered with industry through the NACA advisory committees and, less so, the annual meetings at the Centers. Such ties offered genuine opportunities for input from industry, as well as valuable public relations vehicles for an organization that had to be careful about how it lobbied Congress and the White House. During the space race, NASA did not have to think about currying anyone's favor. Who had time to think about whether NASA was being sensitive to the needs of industry? So, in a sense, the post-Apollo NASA was getting a taste of what it had been like in the NACA years. James Fletcher, when he assumed the leadership of NASA in 1972, was encouraged by the White House to strengthen ties to industry by exploiting the advice

16. James C. Fletcher to George P. Shultz, 30 September 1971, James C. Fletcher Correspondence 1971, folder 4247, HRC.
17. Memorandum for the Record, "Center Directors' Meeting, 10–11 December 1973, Cross Keys Inn, Columbia, MD," 18 December 1973, James C. Fletcher Correspondence 1973, folder 4249, NASA HRC.

of business leaders and by pursuing technological research that had commercial promise for American industry.[18] With NASA's funding under the axe and public support for the space program on the wane, reaching out to industry had obvious benefits. Edward Gray, the Assistant Administrator for Industry Affairs and Technology Utilization, summarized the value of such ties: "Our interest in [aerospace and non-aerospace] companies is to gain their support of NASA in the administration, with Congress, and with the U.S. public. To do this we will have to convince these companies that NASA program objectives, such as development of the shuttle, will benefit them."[19] Fletcher and his deputy, George Low, initiated a series of meetings away from Headquarters, doing the NACA one better by bringing non-aerospace companies into the fold. The feedback from these discussions was generally positive (at least from Edward Gray's perspective), but industry was not always interested in a more commercially oriented Agency. At an October 1973 meeting held at JPL in California, Gray noted that "[a] NASA policy for early dissemination of technology having commercial value to domestic industry did not appear to be of great interest to this group," and executives indicated that "[t]oo much emphasis on today's nonaerospace problems will dry up the well of new technology being developed for the future."[20] The more substantive change in regards to industry, however, was not Fletcher's road show, but awarding R&D contracts to companies for the purpose of applying NASA ideas. This evolved primarily in the ACEE program examined below.

By 1975, the distinction between science-oriented and mission-oriented Centers was all but gone as Headquarters sought to enact a blanket system for approving and allocating projects to the Centers. While project management remained at the Center level, program management resided at Headquarters. New projects, regardless of which Center might have proposed them, would go through the appropriate program's Associate Administrator and then be approved by an Associate Administrator, the Associate Administrator for the Centers, the Deputy Administrator, and the Administrator. All major projects

18. Fletcher to Low, "Luncheon Conversation with Pete Peterson."
19. Edward Z. Gray to A/Administrator and AD/Deputy Administrator [James C. Fletcher and George M. Low], "NASA Relationships with Industry," 16 May 1973, James C. Fletcher Industry Relations, folder 4228, NASA HRC.
20. Edward Z. Gray, Memorandum for the Record, "Discussions with Executives Regarding NASA," undated but referring to 3 October 1973 meeting, James C. Fletcher Industry Relations, folder 4228, NASA HRC. Fletcher had another round of meetings with executives on the East Coast in January 1974: Edward Z. Gray, Memorandum for the Record, "Meeting with Industry Leaders, January 15 and 16, 1974," undated, James C. Fletcher Industry Relations, folder 4228, NASA HRC.

required full plans and analyses with clearly defined work. All projects were to spell out the programmatic, managerial, resource, and schedule "implications." All Centers were to retain some discretionary spending power and carry out some amount of science and "research and technology" in order to remain at the forefront and avoid "technological stagnation." Further, the Centers were to become "centers of excellence," a move intended to eliminate duplication of resources.[21]

Computational Fluid Dynamics

Computational fluid dynamics, or CFD, did not mark the first application of computers to aeronautical research. Through the 1960s, NASA's Centers maintained computer facilities that assisted in the analysis of test data and, relative to CFD, less complex mathematical operations such as linear equations. As was typical for the era, these were usually centralized branches that served many projects, with personnel trained to code problems presented to them by the other branches.[22] Test data, as from wind tunnels and flight testing, also benefited from electronic computers' ability to sift through large amounts of data and reduce it to something meaningful. CFD, however, was a revolutionary jump, an attempt to model the behavior of fluids on a computer. Using a computer to replace a wind tunnel was hardly an intuitive step; fluid dynamics are so irreducibly complex that expecting a computer to compete with experimental apparatus took a leap of faith. One might argue that CFD proponents were acting irrationally, and, indeed, their early efforts were, on the face of it, not terribly successful. Still, the early work served to educate researchers about computational design, especially parallel processing. As computational power increased, CFD began to turn the corner and convince skeptics that it had a place in the lab. Langley, Lewis, and Ames all pursued aspects of CFD. The Ames narrative is of special interest because that Center represented the most contentious bid to use the technology. Competition among the Centers ensured the growth of competing approaches to the mathematical obstacles posed by CFD.

---

21. E. S. Groo, Associate Administrator for Center Operations,, to James C. Fletcher, "Roles and Missions," 19 March 1975 [includes attachment, "Management of Roles and Missions], James C. Fletcher Correspondence 1975, folder 4251, NASA HRC.
22. Christine Darden first began work at Langley in 1967 as a female "computer"; she distinguished herself by coding these mathematical operations for electronic computation. See Christine M. Darden, interview by Robert G. Ferguson, tape recording, Langley, VA, 3 March 2005, NASA HRC.

The principal figure in CFD at Ames was Harvard Lomax, a Stanford graduate in mechanical engineering who joined the Center in 1944. He first worked in the 16-foot wind tunnel and then moved to the theoretical aerodynamics branch, where he honed his mathematical understanding of fluid dynamics. Lomax became convinced of the utility of electronic computers in 1959 and shortly thereafter taught himself how to program. In the early 1960s, he showed how computers could be used to predict fluid flow. One of his standout projects was showing the behavior of blunt-body objects, providing a design tool that complemented H. Julian Allen's earlier work. At a time when computers were commonly seen as adjuncts to wind tunnels and flight testing, Lomax was showing that they could produce valuable data all on their own. Lomax was not alone in his belief in CFD, nor was he the first either for the field in general or at Ames. Lomax was part of a gifted team that included Frank Fuller, Milton Van Dyke (who left for Stanford), and Max Heaslet, but Lomax often provided critical insights, and for this he rose within the theoretical branch.[23]

One factor behind CFD's early support at Ames was strategic positioning within NASA. Ames, to reiterate, began as a West Coast Langley. Since World War II, however, Langley had retained its position as the preeminent Center for aviation-related aerodynamics. Ames risked duplicating Langley's capabilities if it did not seek out new research possibilities. In 1969, Hans Mark arrived to head Ames, replacing H. Julian Allen. Mark was Ames's first outside Director, and he did not have a background in aeronautics. As a physicist with expertise in nuclear science at MIT and the University of California, Berkeley, Mark was aware of the growing use of powerful computers in his own field. Indeed, Ames researchers would later argue that CFD was analogous to the experience of nuclear physics, where neutron transport mechanics and trajectory mechanics modeling reduced the role of live reactor research to merely validating computer modeling.[24] Along with Dean R. Chapman, Mark decided that Ames was going to push CFD, a move that antagonized some researchers. Mark made Lomax head of a newly minted CFD branch (formerly the theoretical branch of the Thermo and Gas Dynamics Division). They began securing ever more

---

23. For a concise biography on Lomax, see Michael R. Adamson, "Harvard Lomax, 1922–1999," *IEEE Annals of the History of Computing* (July–September 2005): 98–102. For early perspectives on the rise of CFD, see Victor L. Peterson, oral interview by Robert G. Ferguson, Los Altos, CA, 17 January 2005, NASA HRC; and Vernon J. Rossow, interview by Robert G. Ferguson, 13 and 18 January 2005, Ames Research Center, NASA HRC.
24. NASA Office of Aeronautics and Space Technology, "The Numerical Aerodynamic Simulator—Description, Status, and Funding Options," December 1981, folder 8740, NASA HRC.

powerful computers, and, in 1970, Mark managed to acquire the Illiac IV, a supercomputer that sought high speeds through parallel processing.[25]

Up to the point of installing the Illiac IV, Ames had a succession of computers. The Center's first computer was an electronic analog computer for icing research. The first digital computer was an IBM 650, used primarily for wind tunnel data reduction. It was also available for researchers to try out their own programs. Lomax ran his first programs on the 650. From there, Ames's central computer facility installed an IBM 704, an IBM 7090, an IBM 7094, an IBM 360, and a CDC 7600. Lomax's theoretical branch consumed as much as a third of the computing time of these machines in the early 1960s, so the branch argued forcefully for new and more powerful equipment.[26] The Illiac IV represented a distinct acquisition, however. The Illiac IV was a joint University of Illinois and Defense Advanced Research Projects Agency (DARPA) project begun in 1964. The custom-designed computer took years to build, and well before it was complete, student protests led the University of Illinois to seek a new location for the computer, as the students were suspicious of the DARPA-funded project. Hans Mark campaigned for and received the controversial system. It was finally installed in 1972, but it was another four years before it was operational. Illiac IV was revolutionary because it attempted to solve problems by using many processors in parallel. This, however, required new software to manage the problem, so Ames had to commit large numbers of researchers to creating novel programs. Though the computer is credited with accomplishing a number of firsts, such as modeling separated flows, its larger impact was on the skill base of the Ames CFD group. The Illiac IV took too long to build and too long to program, and it is remembered as a troubled system, but it helped educate the group about parallel programming (which would evolve into vector processing) and the importance of designing high-performance computing facilities that were easier to program and use.[27]

Computer processing power quickly advanced over the coming decade, but what enabled CFD were the insights that allowed researchers to program very complex mathematical operations. Lomax and his team had to contend with

25. See the Rossow interview.
26. Ames to NASA Headquarters, "Purchase of IBM 704 Computing System," 23 June 1961, Computer Files, folder 8741, NASA HRC.
27. See the Rossow interview for Ames computing history and the role of the Illiac IV. Historian Michael Adamson credits the Illiac IV with being able to "simulate separated flows, airfoil buffeting and buzz, aerodynamic noise, surface pressure fluctuations, and boundary-layer transition." Adamson, "Harvard Lomax." Gina Bari Kolata, "Who Will Build the Next Supercomputer?" *Science* 211 (16 January 1981): 268–269.

two major constraints. First, they could only program problems that computers could solve. This meant learning to reduce differential equations (e.g., the fundamental Navier-Stokes equations that describe fluid dynamics) into algebraic approximations.[28] Creating finite difference techniques that allowed for the simulation of transonic flows was one of the areas in which Harvey Lomax excelled. Second, they could only program problems that a computer could solve within a reasonable amount of time. This meant limiting the scope of the problem. One way they did this was to solve the problem only for a small number of points. They defined a three-dimensional grid, or matrix, within the fluid flow and placed their shape at the center. Ames aerodynamicist Vern Rossow recalled that they had learned some of these techniques from reading unclassified papers on nuclear simulation.[29]

In July 1973, the American Institute of Aeronautics and Astronautics held the First Computational Fluid Dynamics Conference in Palm Springs, California. The event signaled the growing currency of CFD in aerospace research and design. In 1975, Dean Chapman, Hans Mark, and Melvin Pirtle published an article in *Astronautics and Aeronautics* arguing that a major shift toward CFD was under way in their field in spite of contemporary limitations.[30] Friction grew between staunch wind tunnel advocates and CFD's champions. This was not entirely a question of whether CFD had merit. Over time, it became clear to most that CFD was a powerful tool, even if it could not model everything that a tunnel could. At Ames, the tension between CFD and tunnels was exacerbated by how CFD was being deployed. Although some wind tunnel aerodynamicists trained in CFD (some even rotating through Lomax's lab) and might have envisioned a mixed lab (e.g., studying the transonic region using both a tunnel and a powerful computer), computing equipment was given primarily to the CFD camp. Management encouraged, in the words of Vern Rossow, a "combat attitude" between the two groups, though CFD's advance always relied on deep cooperation with the tunnels in order to verify code.

28. "Technically classed as nonlinear, second-order, partial differential equations, [the Navier-Stokes equations] contain over 60 partial-derivative terms when expressed in Cartesian coordinates." Benjamin M. Elson, "New Computers Will Aid Advanced Designs," *Aviation Week and Space Technology* (29 August 1983): 50–57.

29. The Vernon Rossow interview is especially helpful in understanding the mathematical underpinnings of CFD. See also Adamson, "Harvard Lomax."

30. Antony Jameson, "CFD for Aerodynamic Design and Optimization: Its Evolution over the Last Three Decades" (16th AIAA CFD Conference, Orlando, FL, 23–26 June 2003); Dean R. Chapman, Hans Mark, and Melvin W. Pirtle, "Computers vs. Wind Tunnels for Aerodynamic Flow Simulations," *Astronautics and Aeronautics* 13 (April 1975): 22–30, 35.

Combined with a move toward subcontracted wind tunnel operations (rather than having them run by NASA technicians), the trend for wind tunnels was becoming all too clear.[31] By the second half of the 1970s, Ames's grand plans focused squarely on a followup to the Illiac IV, a computing center focused not on a specific machine but on the promise of providing advanced supercomputing capability to researchers nationwide.

Researchers at Langley also responded to the digital revolution. Like Ames, Langley had its own computing facilities and also pursued CFD. The Center had a Star 100 computer in the 1970s, which was Control Data Corporation's first attempt at a vector-processing supercomputer, but Langley pursued CFD in a distinct manner, establishing computing branches as complements (rather than as rivals) to the tunnels. The Center set out in two organizational directions. In 1972, Langley established the Institute for Computer Applications to Science and Engineering (ICASE). ICASE, operated by the Universities Space Research Association (USRA), was more broadly conceived as a scientific computational center meant to encourage collaboration between NASA and academics.[32] Unlike Ames, Langley did not focus on building hardware. Langley also established a number of CFD labs that tended to mirror the division of labor among the tunnels. They had CFD laboratories working in the transonic, high speed, and low speed regimes, as well as aeroelasticity. In the broad scheme of Langley's research, CFD was positioned as a partner to the tunnels, not a replacement. Meanwhile, at ICASE, Langley and visiting researchers advanced alternative mathematical approaches for use in CFD. Some of the principal figures in this included Yousuff Hussaini, who was one of the directors at ICASE; David Gottlieb, a visiting ICASE researcher from Tel-Aviv University; Steven Orszag, a collaborator at MIT; and Thomas Zang, of Langley.[33] In 1983, Ames followed Langley and established a USRA center, the Research Institute for Advanced Computer Science (RIACS), modeled on ICASE.[34]

---

31. See the Rossow interview and the Peterson interview.
32. Institute for Computer Applications in Science and Engineering, "ICASE Semi-Annual Report" (NASA CR-142239, September 1974–February 1975).
33. David Gottlieb and Steven A. Orszag, "Numerical Analysis of Spectral Methods" (NASA Contractor Report CR-157778, 1 June 1977); David Gottlieb, M. Yousuff Hussaini, and S. A. Orszag, "Theory and Applications of Spectral Methods" (NASA Contractor Report CR-185818, 19 December 1983); M. Y. Hussaini and T. A. Zang, "Spectral Methods in Fluid Dynamics" (NASA Contractor Report CR-178103, May 1986).
34. RIACS, Annual Report, October 1997–September 1998, Research Institute for Advanced Computer Science, NASA Ames Research Center, p. 2.

Digital Fly-By-Wire

In a conventional aircraft, mechanical linkages (cables and hydraulic lines) connect the pilot's primary controls (e.g., the wheel/stick and pedals) to the control surfaces. Engineers can design the linkages to amplify a pilot's strength through mechanical advantage or boost pumps, as well as dampen oscillations with springs and pistons. In a conventional aircraft, the pilot is, in every sense, flying the aircraft. Digital fly-by-wire (DFBW) aircraft, in contrast, use electrical signals to transmit a pilot's inputs to the control surfaces *and* interpose a computer between the two. In a sense, the pilot no longer flies the aircraft; rather, the pilot gives roll, pitch, and yaw commands, and the computer decides which combination of control surface changes will achieve the desired result.[35]

The idea of fly-by-wire held a number of attractions for engineers. Aircraft designers could use the computer to actively make the aircraft stable rather than creating a passively stable aerodynamic structure; software replaces aluminum, if you will. A less stable structure may, at first glance, appear to be an undesirable trait, but it greatly expands the design envelope. The problem with unstable designs is that they are difficult, if not impossible, for humans to pilot; humans are limited in the number of inputs they can process in real time. Electronic computers are less constrained; they can process a larger quantity of sensor inputs than humans and transmit precise control signals every few milliseconds.

Thus, fly-by-wire offered the possibility that engineers could design shapes defined by their function and with less regard for their controllability. For combat aircraft, this meant new levels of maneuverability, as there is an inverse relationship between maneuverability and stability. For other aircraft, such as the lifting-body designs that the researchers at the Flight Research Center wrestled with through the 1960s, this meant that the most obstreperous and dangerous of vehicles could be tamed. For all aircraft, fly-by-wire meant that control surfaces could be reduced, optimized, and, in some cases, eliminated. With this development, designers could realize significant savings in weight and drag.

As attractive as the technology seemed, it also held the potential of being a kind of engineer's fantasy. Cables and pulleys, if nothing else, were robust systems, and of all the qualities of a flight control system, robustness is perhaps the most important. A system can be heavy and unforgiving, but at the very least it has to work all the time. Regardless of fly-by-wire's distinct advantages,

35. The definition of "fly-by-wire" has evolved. It originally referred to systems that merely replaced mechanical linkages with electrical linkages. James E. Tomayko, "Computers in Spaceflight: "The NASA Experience" (NASA CR-182505, March 1988), p. 28.

the key to the technology's successful innovation (and adoption) was merely replicating the reliability of systems it replaced.

There is an element of technological determinism in the decision to pursue fly-by-wire at NASA. This was a technology that could only progress with the advent of electronic computers sufficiently small and reliable to operate in an aircraft. NASA was uniquely positioned in this respect because it had taken the lead in developing just such a computer for the Apollo Lunar Exploration Module. Thus, the time was ripe, and NASA aeronautical engineers enjoyed access to this new technology. On the other hand, the research was not necessarily a logical unfolding of technological change. The computers of the era were expensive and temperamental. By attempting to create a reliable system out of unreliable hardware, NASA was making a very difficult task for itself. Indeed, even with the more advanced computers of the second phase of the project, NASA found that merely keeping the computers in working order was a feat. Furthermore, NASA's choice of a digital system was not necessarily the "right" one. The U.S. Air Force, simultaneously, was also examining fly-by-wire systems but was doing so with analog systems (electrical equipment that operated by signal manipulation rather than the discrete on/off switches of a digital system). Finally, engineers could have chosen to implement digital control in a number of ways, but NASA's researchers took a high-risk approach and applied it to the pilot's primary controls. Indeed, the adoption of DFBW in commercial aircraft took a different route, first with widespread application of digital controls to jet engines and then to the pilot's primary controls.

Engineers at the Flight Research Center initiated NASA's DFBW program after casting about for a new project in the late 1960s. Led by Melvin E. Burke, Calvin R. Jarvis, Dwain A. Deets, and Kenneth J. Szalai, the group was already familiar with fly-by-wire technology. Over the previous decade, the FRC had experimented with reaction control systems for the X-15; partial fly-by-wire control in the lifting-body program; and the dangerous but challenging Lunar Landing Research Vehicle, a fly-by-wire contraption that let astronauts practice simulated Moon landings.[36] The decision to pursue a digital system, and one that governed the pilot's primary controls, was partly strategic and partly rhetorical. The U.S. Air Force was already funding its own fly-by-wire research, notably an F-4 Phantom refitted with a three-axis analog computer control.

36. James E. Tomayko, *Computers Take Flight: A History of NASA's Pioneering Digital Fly-By-Wire Project* (Washington, DC: NASA SP-4224, 2000), pp. 21–22; Gene J. Matranga, Wayne C. Ottinger, Calvin R. Jarvis, and Christian D. Gelzer, *Unconventional, Contrary and Ugly: The Lunar Landing Research Vehicle*, Monographs in Aerospace History, no. 35 (Washington, DC: NASA SP-2004-4535, 2006).

There was no point in NASA's attempting the same thing, and attempting anything less would have been too timid. NASA could carve out its own niche with digital, three-axis control and in doing so stake out the highest-risk position. More than strategy, if the FRC's gambit proved successful, NASA would have demonstrated the capabilities of DFBW in its most demanding task. Thus, DFBW was not simply a technical choice, but one delimited by a competitive research environment and an often-skeptical aircraft industry. At its inception, it had to open Headquarters' purse strings; and at its close, win the confidence of industry.[37]

Interestingly, the technical outlines of the program took shape only when two members of the FRC team, Melvin Burke and Calvin Jarvis, went to Headquarters and presented their idea to then–Deputy Associate Administrator Neil Armstrong. Armstrong suggested that they look at using the Apollo program's digital computers and that they contact its designer, the Draper Laboratory. With funding approved, Burke's team did just that. In a single move, the DFBW project had a computer and the support of one of the nation's most technically competent communities. The DFBW project was not only piggy-backing off of a substantial Apollo-era investment but also had the advantage of the Draper Lab's long involvement in military guidance systems, notably the Navy's Polaris missile system. For the FRC, integrating and testing different pieces of equipment for this project meant integrating vastly different specialist communities.[38]

From the start, the DFBW project was designed as a series of phases. Officially, phase 1 was to give NASA experience in DFBW, show that "dissimilar redundancy" could work (i.e., the digital primary and analog backup), and show that the airplane could be controlled by a software program.[39] Phase 1 was not sufficient to prove that the technology was ready for commercial aircraft, but it was enough to create momentum for further research funding, and it established the Center's competency in the technology. Subsequent research plans evolved with NASA's budget, with the Agency's overarching priorities, and with the knowledge gained from phase 1. In the project's most ambitious plans, researchers such as Ken Szalai considered exploring the technology on a highly unstable aircraft, a goal that would only be realized in different projects in the 1980s. As it turned out, phase 2 emphasized system reliability with the use of redundant computers and adaptive flight controls.[40]

37. Tomayko, *Computers Take Flight*, pp. 23, 29–30.

38. Ibid., chap. 2.

39. Ibid., p. 69.

40. Ibid., pp. 32–33, 82–87.

Back at the FRC, the group acquired a set of ex-Navy F-8 aircraft, resurrected from the boneyard for a second life of experimentation. One of these F-8s would fly the Center's supercritical airfoil experiments. For the DFBW program, technicians removed the F-8's mechanical controls and set about replacing them with a digital system and an analog backup (thus, even the backup was fly-by-wire). Another F-8, the "Iron Bird," also had its systems replaced with the same digital and analog electronics. Engineers used this grounded aircraft to validate hardware and software before testing it in the flight-ready F-8.[41]

As noted, the DFBW program's major hurdle was creating a system as robust as conventionally controlled aircraft. They sought a reliability level of 99.99999 percent (as opposed to a historical figure of 99.999565 for conventional, commercial aircraft).[42] Some of this reliability would have to be in the hardware, such as using triply redundant actuators and the robust Apollo computer, but the true Achilles heel was the flight control software. It had to respond in a way that was predictable for the pilot, lest the pilot begin to overcorrect and make the situation worse.[43] This had to be done in all three axes, each one requiring its own set of flight laws. Even assuming that the engineers had these laws described perfectly from the start, they had to write a program around the equations that would flawlessly process sensor inputs and consistently produce the expected output. The opportunities for error in such a complex undertaking were manifold.

Teams at both Draper and the FRC tackled the software problem. Jarvis, Deets, and Szalai wrote the software specifications for the different axes (pitch, roll, and yaw). Working with the FRC's input, Draper's software engineers built the flight control laws into the Apollo software. They retained some 60 percent of the original Apollo software, such as executive code and diagnostic tests. When they deemed a version of the software completed, Draper wrote the code into the Apollo computer's core rope memory. The rope memory could not be rewritten, so both Draper and FRC teams checked and rechecked their work for errors.[44]

Buttressing the reliability of the complex software was a hardware design with built-in error checking and backups. The single Apollo computer gave its output to two digital-to-analog converters, the second one monitoring the first.

---

41. Ibid., pp. 32–33, 48–49.
42. Ibid., p. 46.
43. Tomayko describes, for example, how engineers had to build a "deadband" region into control movements. Pilots were accustomed to mechanical systems in which small movements of the control column did nothing initially because of slack in the cables. Tomayko, *Computers Take Flight,* p. 50.
44. Ibid., pp. 43, 49, 54.

(NASA image E-24741)

Figure 4.2. Digital fly-by-wire electronics installed in Dryden's F-8 aircraft, 1971.

If the computer detected a fault, or if the outputs did not agree, the system fell back on an analog control system. This procedure was not trivial because the analog system had to take immediate control of the aircraft without upsetting it. To manage this, the digital system constantly synchronized with the analog, giving the latter up-to-date sensor and stick information. At the far end of the process, the hydraulic actuators themselves had backup modes.[45]

Phase 1 testing lasted from May 1972 to November 1973. In addition to validating the technology, the F-8 flew with a side-stick attached to the backup control system. General Dynamics, at this time, was developing its YF-16, a highly maneuverable fighter prototype that used an analog fly-by-wire system. It was a good opportunity to use the F-8s as a test bed; after installing the prototype device, NASA flew the side-stick successfully on six test flights.[46]

In parallel with phase 1, Jarvis and his team (Burke had left for a position at Headquarters) began planning the follow-on research. A primary goal of phase 2 was the use of redundant, commercially available computers. This was

---

45. Ibid., pp. 54–55, 61–64, 66.
46. Ibid., pp. 79–82.

essential to proving that the technology was ready for commercial aircraft rather than a highly specialized design for research and military aircraft. Choosing a new computer, however, meant that the reliability work that was done for the Apollo computer (much of which was completed before the DFBW program began) would have to be repeated. This was new hardware in a new configuration using new software. Of course, phase 2 appropriated some of the earlier design principles and processes, but it was akin to starting from scratch. Indeed, these development hurdles contributed to phase 2's lengthy run of 12 years.[47]

As in phase 1, the space side of NASA played a large role in phase 2. In the search for the best computer, NASA had a number of options recently available from commercial vendors. Cost, weight, volume, power consumption, and, most importantly, computational capability (speed and memory) differentiated the contenders, but in the end, the team decided to use the same computer as the forthcoming Space Shuttle. Managers from the Space Shuttle Program had taken an interest in the DFBW project since the Shuttle would be, like all previous spacecraft, fly-by-wire. Critically, however, the Shuttle would need to fly like a glider during reentry. Shuttle program managers extended an offer to the DFBW program: purchase the same IBM AP-101 computers and they would receive $1 million in funding from the Shuttle Program. This turned out to be a bargain for the space program.[48]

The FRC refitted the same F-8 aircraft with the IBM computers while Draper again performed its coding function. This time, Draper designed the system so that the three AP-101s synchronized with each other every few milliseconds, taking inputs, performing diagnostic tests, and sending outputs to the actuators. Researchers at Langley had an expanded role in phase 2; they contributed advanced control laws that, when turned on, optimized all of the control surfaces for a given pilot input. In essence, the pilot told the computer where to go, and the computer decided how best to use the available control surfaces in concert.[49] Beyond validating the redundant computer system, phase 2 explored adaptive control, remote augmentation (i.e., using a computer on the ground to make changes in flight control parameters in flight), and experiments on the system's robustness (e.g., learning how to safely reduce sensor inputs).[50]

The first flight in phase 2 took place in 1976, and over the course of the next nine years, the F-8 flew 211 flights. At its height, the FRC had some 50 people working on the project, a large project for aeronautics, but a

47. Ibid., p. 85.

48. Ibid., p. 93.

49. Ibid., p. 109.

50. Ibid., pp. 114–116.

comparatively small one for the space program.[51] The Shuttle Program, which had vastly greater software requirements than the F-8, still benefited from the DFBW experience and code. Shuttle pilots flew the F-8 using the Shuttle's backup flight software in 1977, while engineers worked on the problem of pilot-induced oscillations.[52] Follow-on programs to the DFBW included the Digital Electronic Engine Controls (DEEC) program and Highly Integrated Digital Electronic Control (HIDEC) programs.[53] Digital fly-by-wire found application in unusual shapes that are, by humans alone, impossible (or nearly impossible) to pilot. The stealth shapes of the B-2 bomber and the F-117 are made controllable by fly-by-wire, as was the highly unconventional forward-swept-wing aircraft, the NASA-USAF X-29.

Energy Efficiency

In 1973, OPEC, the Organization of the Petroleum Exporting Countries, began its oil embargo in retaliation for western support for Israel during the Yom Kippur War. Through the year and into 1974, the price of oil rose steeply, with substantial repercussions for the aviation industry. Airlines saw profit margins erode and faced jet fuel shortages while the economics for fuel-hungry aircraft like the Anglo-French Concorde became untenable without government subsidy. With assumptions about inexpensive jet fuel shattered, the logic underpinning aircraft and jet engine design shifted. At NASA, Albert Braslow, the assistant head for the Advanced Transport Technology Office, noted in late 1974, "Fuel is now a design parameter."[54]

Of course, energy efficiency was nothing new for NASA, or its predecessor, the NACA. Indeed, much of its earlier work could have been construed as one form or another of efficiency research since any increase in efficiency had immediate consequences for range, payload, and performance. For example, Whitcomb's long-term study of drag reduction, whether at supersonic or subsonic speeds, was, in effect, efficiency research. So by the time of the oil embargo, NASA already had programs that were immediately applicable to the problem at hand. Researchers also knew what they would do to attack the issue; in 1974, *Aviation Week & Space Technology* reported that NASA's near-term

51. Ibid., pp. 93, 99, 123.

52. Ibid., For a description of the F-8's contribution to the Space Shuttle, see James Tomayko's "Computers in Spaceflight: The NASA Experience," chap. 4.

53. Tomayko, *Computers Take Flight*, p. 133.

54. Warren C. Wetmore, "Fuel Outlook Dictating Technical Transport Research," *Aviation Week & Space Technology* (28 October 1974): 52–63.

(within one decade) solutions included "supercritical aerodynamics, composite materials, advanced propulsion and avionics and active controls."[55]

In February 1975, George Low, NASA's Deputy Administrator, directed Alan Lovelace, the Associate Administrator for Aeronautics and Space Technology, to initiate an aeronautical energy-conservation program. "Conservation of energy is a matter of high national importance," he wrote.[56] George Low was likely responding to requests from the Senate for NASA to do something about the oil shortage.[57] Within a month, Lovelace had approval for an Advisory Board for Aircraft Fuel Conservation Technology composed of a NASA team and representatives from DOD, DOT, and the FAA.[58] Low insisted that they were to "develop a program within existing NASA resources." When it appeared that Lovelace's group plans were becoming more expansive, Low noted that "…it was never my intention that the aircraft fuel conservation program be carried out on top of anything else we are doing in aeronautics."[59] Eventually, the Advisory Board arrived at a short list of options that, not surprisingly, echoed NASA's earlier orientation:

- Engine Component Improvement
- Fuel Conservative Engine
- Fuel Conservative Transport (aerodynamic design, active controls)
- Turboprops
- Laminar Flow Control
- Composite Primary Structures[60]

55. Ibid.
56. George M. Low to Alan M. Lovelace (AA for Aeronautics and Space Technology), "Aeronautical Energy Conservation Program," 1 February 1975, folder 18273, NASA HRC.
57. Robert W. Leonard, the manager of the ACEE program at Langley, reported in 1978 that "[t]he great impact of fuel costs resulted in requests from the Senate for NASA to identify new technology to promote fuel conservation." Robert Graves, "Research Focuses on Fuel Efficiency for Planes," *Daily Press* (Newport News, VA) (19 February 1978), folder 18273, NASA HRC. See also James J. Kramer, "Planning a New Era in Air Transport Efficiency," *Astronautics and Aeronautics* (July/August 1978): 26–28. James Kramer was Associate Administrator of OAST when he wrote the article.
58. James C. Fletcher to Lovelace, "Approval for Establishment of Advisory Board for Aircraft Fuel Conservation Technology," 13 March 1975, folder 18273, NASA HRC.
59. George M. Low to Alan Lovelace, "Comments on Preliminary Draft Report of the Aircraft Fuel Conservation Technology Task Force," 30 April 1975; Advisory Board on Aircraft Fuel Conservation Technology, Task Force Report, 10 September 1975, OAST, folder 18273, NASA HRC.
60. NASA News, "Senate Group Briefed on Aircraft Fuel-Saving Effort," release no. 75-252, 10 September 1975, folder 18273, NASA HRC.

Laminar flow control had been mentioned a year earlier in the *Aviation Week* article. The only new addition was turboprops, and jet engine propulsion was divided into a near-term effort that would seek incremental improvements in engine components and a longer-term effort at new designs. NASA briefed the Senate Committee on Aeronautical and Space Sciences in September 1975.[61]

Planning continued through 1976 following the same research outline established by the Advisory Board, but with one significant change. Where George Low envisioned an effort operating within NASA's existing programs, the Aircraft Energy Efficiency Program, as it came to be known, operated primarily through a set of subcontracts with the airframe and engine manufacturers. While the Senate committee encouraged NASA to work closely with industry (something that suggested the outright subsidy of corporate research), there were also pragmatic reasons for creating this arrangement. The program included some technologies that, so far as NASA was concerned, had already been validated in the laboratory, but industry was not yet convinced. Performing more laboratory work was not necessarily going to win over the private sector.[62] By being paid to apply the technology to existing aircraft and test it operationally, the manufacturers would gain confidence while generating reliable cost/performance estimates. Moreover, especially in regard to the jet engine programs, NASA relied greatly on operational equipment for generating ideas about efficiency improvements. That is, laboratory work was not necessarily where NASA's researchers would find their best answers.

By 1978, five years after the initial oil price shocks, the logic for the ACEE program had expanded: ACEE represented an answer to changing societal priorities and a defense against foreign aircraft manufacturers. Arguing that the aviation industry needed to get in step with widespread sentiment about technology, D. William Conner of Langley wrote in mid-1978 that "the industry must address the concerns of the users *and* the public. The traveler, for example is concerned with safety, cost, frequency of service, total trip time, and comfort. The *public*, however, is concerned with conservation of resources, minimization of environmental impact, and having a system to meet the needs of the entire country."[63] The Newport News *Daily Press* (which closely followed aviation developments at nearby Langley) reported that the ACEE "would

61. Advisory Board on Aircraft Fuel Conservation Technology, Task Force Report, 10 September 1975.

62. NASA's work on supercritical airfoils was, by this time, largely complete, yet manufacturers were not implementing the technology and did not have the design tools to do so easily. See Thomas Grubisich, "Fuel-Saver in Wings," *Washington Post* (11 July 1974): C1.

63. D. William Conner, "CTOL Concepts and Technology Development," *Astronautics and Aeronautics* (July/August 1978): 29–37. Conner was head of Langley's Systems Analysis Branch at the time. It

not only benefit when oil-based fuel becomes scarce, it could also halt or slow the threatened large-scale invasion of the world air transport market by foreign manufacturers and help stave off future massive trade deficits.... The [Airbus] aircraft is the first two-engine wide-bodied jet to fly commercially in the U.S.... The Airbus is deemed quieter, smaller and more fuel efficient than its U.S. competitors."[64]

In the category of aerodynamics, NASA already had a technology ready to sell: Whitcomb's supercritical wings. Whitcomb considered this research largely finished even by the time his experimental wings began flying on an F-8 at the FRC. The ACEE program sought to encourage the use of the supercritical airfoils by creating computer models that would ease the design of such wings and expand the range of useful configurations. Following the supercritical airfoil work, Whitcomb investigated the problem of induced drag on wingtips where the air swirls off the end and back over the top of the wing. He proposed placing a couple of small vertical wings within this stream in order to convert the energy into a forward component, thus reducing drag. A longer wing would provide a similar benefit, but the additional moment (weight multiplied by the distance from the aircraft centerline) is greater than for a winglet of similar aerodynamic performance. Whitcomb's team ran tests in the 8-Foot Transonic Pressure Tunnel from 1974 to 1976 and conducted flight tests on a converted KC-135 three years later. The ACEE program picked up winglets from the beginning; they were an attractive technology, in part because manufacturers could retrofit existing designs (whereas supercritical airfoils would have to wait for entirely new designs). NASA contracted with McDonnell Douglas to design, fabricate, and test a set of winglets on a leased Continental Airlines DC-10. Flight tests, interestingly, were conducted from the Douglas Long Beach facility rather than at NASA's FRC.[65]

The ACEE program also encouraged the use of active controls (as earlier conceived in the DFBW program) that would allow the reduction in the size of control surfaces. ACEE even incorporated funding for the development of fault-tolerant computers to be used in active control systems.[66] The three U.S. airframe manufacturers at the time (Boeing, McDonnell Douglas, and Lockheed) each received research contracts, beginning in 1977, with the

---

is interesting to note that even as Conner urged the industry to be mindful of public sentiment, his article included images of an extremely large *nuclear*-powered cargo seaplane.

64. Graves, "Research Focuses on Fuel Efficiency for Planes."

65. Chambers, *Concept to Reality*, pp. 35–44.

66. Robert W. Leonard, "Airframes and Aerodynamics," *Astronautics and Aeronautics* (July/August 1978): 38–46.

government reimbursing 90 percent of the manufacturers' costs.[67] Lockheed, for example, fit an L-1011 with active controls such that it could control its maneuvering load, suppress gusts, and suppress elastic bending.[68]

Laminar flow control (LFC) was another part of the aerodynamic portion of the ACEE. LFC was achieved by creating suction across the top of the airfoil (i.e., creating a vacuum system within the airfoil that drew air in through many small holes). An effective LFC system eliminates turbulence, which causes drag, and creates a smooth flow over the entire wing. From an aerodynamicist's point of view, LFC was theoretically ideal, but it had significant mechanical drawbacks. In the early 1960s, the U.S. Air Force tested the idea on two Douglas Destroyers (X-21A program) and found that they could achieve laminar flow over three-quarters of the wing, but also that the vacuum holes clogged. The challenge for NASA was not to prove that LFC worked, but to find a mechanical system that kept the wing clean without, of course, adding too much weight and cost.[69] Lockheed and McDonnell Douglas both produced LFC test articles.

Composite structures were not new to aircraft design, but the kinds of composites in which NASA was interested, carbon-fiber composites, were restricted to cutting-edge military aircraft where cost and safety were subordinate to performance. In similar fashion to the aerodynamic program, the ACEE project included funding for the testing of composite structures spread across Boeing, McDonnell Douglas, and Lockheed. They were each to build and test prototypes for primary and secondary structures. The primary structures included load-carrying stabilizers, and secondary structures included control surfaces. This effort attempted to familiarize the manufacturers with the fabrication of composites and gain real-world data on performance, maintenance, and cost. Langley also studied the durability of composites, especially in resisting UV damage, temperature, and moisture damage. As with the supercritical wing design programs, Langley examined different programs for assisting the manufacturers in the design of composite parts and subassemblies. Of particular concern to the program, however, was reducing manufacturing costs. This involved finding an appropriate balance between design simplicity and weight reduction.

---

67. NASA News, "Boeing Awarded Second-Phase Contract," release no. 78-170, 1 November 1978, folder 11455, NASA HRC. See also Jeffrey M. Lenorovitz, "Douglas Speeds Energy-Efficient Study," *Aviation Week & Space Technology* (26 March 1979): 27–28.
68. "Energy Efficient Transport Program," presentation photograph, NASA HQ RA78 1140(3), 25 January 1978, Aircraft Energy Efficiency Program, folder 11455, NASA HRC.
69. Leonard, "Airframes and Aerodynamics," pp. 38–46. See also Dennis R. Jenkins, Tony Landis, and Jay Miller, *American X-Vehicles: An Inventory–X-1 to X-50* (Washington, DC: NASA, 2003), p. 28.

Mechanization was considered an attractive means for achieving production cost reductions.[70] During the program, the FAA certified different composite structures: a rudder in May 1976, an elevator in January 1980, and an aileron in September 1981. The manufacturers produced and flew a small sample of each, gaining valuable durability and maintenance data, including life-cycle costs.[71]

Lewis managed the propulsion portion of the ACEE program. Lewis's contribution to the ACEE represented the Center's return to air-breathing engines after having spent the 1960s devoted to spacecraft propulsion. With the end of Apollo, Lewis saw a devastating outflow of researchers. Air-breathing engine work was part of its bid to regain momentum. The advanced turboprop initiative began earlier in the decade and was the Lab's highest-risk answer to the efficiency challenge. Unfortunately for Lewis, the idea was put on hold and did not become an active part of the ACEE program until the 1980s. In the meantime, Lewis worked on the two less risky approaches: engine component improvement and the energy-efficient engine.

The Engine Component Improvement (ECI) Program sought incremental gains in existing jet engines through operational performance analysis and inspection of component degradation. The Energy Efficient Engine Program sought to establish a baseline for a new generation of engines. As with the aerodynamics studies, NASA worked closely with the two major manufacturers: General Electric (GE) and Pratt & Whitney. In this case, however, the problem was not convincing the manufacturers to adopt a Lewis idea, but to learn from operational engines. The engines *were* the laboratories. NASA also brought in members of the airline industry in order to examine changes in parts, operations, and maintenance procedures. This was done to get information from the users and to examine the market feasibility of specific changes. The question was not always whether an engine could be made more efficient, but whether such a change would gain traction among the airlines.[72] The ECI Program concluded in 1981.[73]

The second portion of the program sought to provide manufacturers with new technologies that would increase efficiency by target amounts. Specific fuel consumption was to drop 12 percent while direct operating costs were to drop 5 percent. Again, research contracts went to GE and Pratt & Whitney. This effort was to produce components, not a prototype engine, which was an

---

70. Leonard, "Airframes and Aerodynamics," pp. 38–46.
71. "ACEE Status," presentation photograph, NASA HQ RJ82 499(3), 15 January 1982, Aircraft Energy Efficiency Program, folder 11455, NASA HRC.
72. Donald L Nored, "Propulsion," *Astronautics and Aeronautics* (July/August 1978): 47–54, 119. Nored managed the Energy Conservative Engines Office at Lewis at the time he wrote this article.
73. Rumerman, *NASA Historical Data Book*, vol. 6, p. 189.

interesting distinction because building better components took the manufacturers a considerable distance toward new engines. The first set of contracts was let in early 1978 and covered $83.8 million for GE and $80.4 million for Pratt over five years. Both GE and Pratt received cost-reimbursement contracts that covered 90 percent of the research costs (with research performed at the manufacturers' facilities).[74] Ultimately, GE did produce a prototype, one with a 13.2 percent improvement in specific fuel consumption, though NASA expected further improvements in the manufacturers' forthcoming designs. The program concluded in 1984.[75]

Most of the ACEE programs were ultimately successful in seeing the underlying technology used by manufacturers and airlines, though in some cases this took more time than NASA expected. Engine improvements filtered quickly into new designs, as did small increases in composite structures. The adoption of supercritical wings was spotty, at least until the 1990s, and large-scale composite use waited until the Boeing 787, which first flew in 2009. The Advanced Turboprop Project (covered in the following chapter), did not take hold.

For Headquarters, the ACEE was really the marquee aeronautics project of the decade. Despite the fact that it represented a fraction of the aeronautics budget, despite the fact that this program was cobbled together and included technologies that had already been proven by NASA researchers (as in the case of the supercritical wing), the ACEE was the project that the Agency talked about. In a March 1977 meeting with GE executives, NASA Administrator James Fletcher discussed a number of research initiatives and technological spinoffs, but the only one from aeronautics that he mentioned was the ACEE.[76]

Vertical/Short Takeoff and Landing

Vertical/short takeoff and landing (V/STOL) research encompasses a variety of technologies that attempt to eliminate or reduce the need for a traditional runway. Runways are not always where one wants them, so designers have long imagined ways in which aircraft could become airborne in a confined area. Helicopters, of course, answer the need for vertical takeoff and landing

---

74. NASA News, "General Electric Gets Fuel Efficiency Contract Award," release no. 78-54, 4 April 1978, folder 11455, NASA HRC; NASA News, "Pratt & Whitney Gets NASA Contract to Improve Aircraft Fuel Economy," release no. 78-74, 17 May 1978, folder 11455, NASA HRC.
75. "NASA Sees 27 Percent Fuel Savings from EEE Program," *Defense Daily* (20 January 1984): 101.
76. Meeting Record, "Discussion of Speech Outline and Charts To Be Used for Dr. Fletcher's Talk to the GE Executives on March 23, 1977," Fletcher–NASA–Industry Relations, folder 4228, NASA HRC.

and have been in practical operation since the late 1930s and in serial production since the 1940s with the Sikorsky R-4.[77] Most helicopters, however, trade vertical maneuverability for speed and range because of the complex aerodynamic and structural difficulties that arise as a helicopter moves faster horizontally. The problem is that on one side of the helicopter the blades are advancing in the same direction as the helicopter, while on the other side they are retreating. Left unaccounted for, the advancing side generates much more lift than the retreating side to the point that the retreating side ultimately stalls. Hinging mechanisms adjust for this disparity in lift, but only up to a certain point. Added to this, most helicopter blades are relatively flexible, so at high horizontal speeds, the aerodynamic effects cause unstable loading and dangerous oscillations.[78]

There have been scores of imagined solutions to the V/STOL challenge, only a few of which have made it to the prototype stage.[79] There is no easy categorization of these prototypes. Early attempts focused on creating pure VTOL aircraft, notable examples being the "tail-sitters" that emerged in the 1950s (the Ryan X-13, the Convair XFY-1, and the Lockheed XFV-1). These aircraft were oriented in a straight-up position at takeoff and landing but transitioned to level flight for cruise, at least in theory.[80] There also has been a continuing series of attempts to make helicopters into better-performing cruise vehicles. The Sikorsky S-69, for example, employed rigid, contra-rotating blades, and the NASA X-wing (discussed later) sought to use a small, stoppable X-shaped rotor that would become a wing in horizontal flight.[81] Another strand has been the

77. Jay P. Spenser, *Whirlybirds: A History of the U.S. Helicopter Pioneers* (Seattle: University of Washington Press, 1998), chaps 1 and 2.
78. J. Gordon Leishman, *Principles of Helicopter Aerodynamics* (New York, NY: Cambridge University Press, 2006), pp. 55–56.
79. W. P. Nelms and S. B. Anderson, "V/STOL Concepts in the United States—Past, Present and Future," NASA TM-85938, April 1984, NASA Ames Research Center. The range of V/STOL possibilities is provided in a 1960s-era diagram from McDonnell Aircraft showing some 62 classes, design studies, and prototypes. Reprinted in Martin D. Maisel, Demo J. Giulianetti, and Daniel C. Dugan, *The History of the XV-15 Tilt Rotor Research Aircraft, From Concept to Flight* (Washington, DC NASA SP-2000-4517, 2000), p. 2.
80. Stephen Wilkinson, "Going Vertical," *Air and Space Magazine* (October/November 1996): 4. On the X-13, see Jenkins et al., *American X-Vehicles*, p. 19. For a lucid explanation of the increased energy requirements of V/STOL flight as well as the options available to designers, see A. C. Adler, "Vertical Takeoff," *International Science and Technology* (December 1965): 50–58.
81. On the Sikorsky S-69, see Spenser, *Whirlybirds*, chap. 2. Sikorsky has returned to rigid, coaxial rotors (two stacked rotors turning in opposite directions) in its recent X-2 demonstrator.

development of vehicles that reorient their thrust in order to transition between vertical and horizontal flight. The well-known Hawker Siddeley Harrier fits this description, as well as the tilt-rotor (discussed below) and a handful of prototypes, not all of which successfully managed the transition between the two flight regimes. Finally, there is the class of purely STOL aircraft, aircraft that appear conventional but employ a variety of techniques to increase lift and thus dramatically decrease takeoff and landing distances.

In spite of the very different solutions proposed to address the challenge of V/STOL flight, most of these vehicles share similar obstacles. Because these vehicles must carry equipment that operates across a wider spectrum of flight regimes, they are generally heavier than their conventional counterparts. Additionally, because of the dangers of asymmetric lift that might result from an engine failure, most of these vehicles incorporated complex gearing or ducting arrangements to provide redundancy. These precautions also add weight. Finally, all of the V/STOL vehicles were beset with special control issues, especially during the transition between vertical and horizontal flight and at slow speeds in which conventional controls (e.g., the vertical and horizontal stabilizers) have a reduced effect.[82]

The large number of different V/STOL programs in the United States would not have been possible without funding from the military. Table 4.2 captures most of the vehicles built up to the 1970s. As can be seen, both Ames and Langley either cooperated with the military in vehicle testing or, at some point, received military test vehicles for their own research purposes. Langley was also involved in early testing of the Harrier's ancestor, the Hawker Siddeley P-1127.[83] As is obvious from the list of prototypes, the military had a strong interest in seeing a successful V/STOL aircraft, but successive designs presented intractable aerodynamic and control problems. The researchers at Langley and Ames grew increasingly familiar with these problems and, since the mid-1950s, had been conducting their own research in the area. Prior to

82. Adler, "Vertical Takeoff," pp. 50–58; NASA Langley Research Center, "VTOL and STOL Technology in Review," *Astronautics and Aeronautics* (September 1968): 56–67.

83. The table does not include all of NASA's V/STOL research. It omits testing done on the North American OV-10A, as well as the P-1127 (with the U.K.), the DO3 (with Germany), and the Breguet 941 (with France). NASA press release no. 68-194, "XC-142 VTOL Test Flights," 10 November 1968, STOL, folder 11726, NASA HRC; Robert C. Seamans, Jr., to John S. Foster, 15 July 1966, STOL, folder 11726, NASA HRC; William S. Aiken, Jr., to RX/Director, Advanced Concepts & Mission Division, "Experimental STOL and V/STOL Aircraft," 13 August 1971, STOL, folder 11726, NASA HRC; Hal Taylor, "NASA Expanding STOL, V/STOL Effort," *Aerospace Technology* (17 June 1968): 16–17.

Table 4.2. U.S. V/STOL Research Aircraft, 1954–1971

Design	Mftr.	Type and Description	First Flight	Supporting Organization
XFY-1	Convair	VTOL: Propeller tail sitter	1954	Navy
XFV-1	Lockheed	VTOL: Propeller tail sitter	1954	Navy
XV-3	Bell Heli.	V/STOL: Tilt-rotor	1955	Air Force (Ames)
X-13	Ryan	V/STOL: Jet lift	1957	Air Force
VZ-2	Vertol	V/STOL: Tilt-wing	1958	Army (Langley)
VZ-3	Ryan	V/STOL: Deflected slipstream	1958	Army (Ames)
VZ-4	Doak	V/STOL: Tilt-duct	1958	Army (Langley)
X-14	Bell Aero	V/STOL: Jet lift	1958	Air Force (Ames)
VZ-9	Avro	V/STOL: Peripheral jet lift	1959	Army, Air Force
C-134	Stroukoff	STOL: Boundary layer control flap	1959	Air Force (Ames)
X-18	Hiller	V/STOL: Tilt-wing	1960	Air Force
X-100	Curtiss-Wright	V/STOL: Tilt-propeller	1960	Air Force
XV-1	McDonnell	V/STOL: Compound helicopter	1960	Army
C-130B	Lockheed	STOL: Boundary layer control flap	1962	Air Force (Ames)
XV-4A	Lockheed	V/STOL: Augmented jet lift	1962	Army
XV-5A	GE-Ryan	V/STOL: Fan-in-wing	1964	Army
XV-9A	Hughes	V/STOL: Hot cycle rotor	1964	Army
XC-142	LTV	V/STOL: Tilt-wing	1964	Air Force (Langley)
X-19	Curtiss-Wright	V/STOL: Tilt-propeller	1965	Air Force
X-22	Bell Aero	V/STOL: Tilt-duct	1967	Navy, Air Force, NASA
XV-5B	GE-Ryan	V/STOL: Fan-in-wing	1968	Ames
XV-4B	Lockheed	V/STOL: Jet lift	1969	Air Force

Source: William S. Aiken, Jr., to RX/Director, Advanced Concepts & Mission Division, "Experimental STOL and V/STOL Aircraft," 13 August 1971, STOL, folder 11726, NASA HRC.

the 1970s, however, NASA had supported only one vehicle (the XV-5B), a rebuilt military prototype.[84] By the late 1960s, two NASA programs began to emerge that would see the Agency take a more commanding role: the tilt rotor and STOL. It is worth noting that while technical innovations, such as lighter weight jet turbines, suggested to NASA researchers that V/STOL aircraft were now possible, external factors also appeared to favor the development of V/STOL: helicopter use in Vietnam and rising air traffic congestion on the Eastern seaboard.

84. Robert C. Seamans, Jr., to John S. Foster, 30 November 1966, STOL, folder 11726, NASA HRC.

In the dense jungles of Vietnam, rotorcraft had become a revolutionary addition to modern warfare. Though employed in the Korean War, where they served a supporting role, helicopters became central to fighting in Southeast Asia. Bell's UH-1 was the all-purpose transport vehicle of the Vietnam War, often equipped with door-mounted machine guns. Bell went further and created a slender gunship, the AH-1 Cobra, around the UH-1's mechanical innards.[85] In addition to Bell's helicopters, both Sikorsky and Boeing (which had purchased the Vertol Corporation in 1960) were producing large military transport helicopters. In this context, the advantages of a practical tilt-rotor for medium-range troop transport were obvious.

Meanwhile, in the crowded northeast air corridor stretching from Boston to Washington, DC, FAA managers and airline executives began to grow concerned about the delays and costs incurred from airport traffic. Furthermore, the prospect of adding new airport capacity appeared remote. Communities were increasingly vocal about the rise in jet noise. STOL technology, which to this point had held little utility for commercial passenger aircraft, offered a theoretical solution. A parallel transportation system of STOL airports, or STOLports, and STOL flightpaths could reduce the traffic burden on conventional runways and flight patterns. The key to making STOL technology economical, therefore, was creating a new transportation system.

The Tilt-Rotor

A tilt-rotor aircraft takes off like a helicopter but flies like a fixed-wing aircraft. It does so by tilting two rotor blades placed at its wingtips. For takeoff and landing, the rotors are pointing upward; for level flight, the rotors tilt forward, becoming, in effect, oversized propellers. Because of the size of the rotors, the aircraft cannot land with them facing forward; the aircraft must transition back to being a helicopter first. The advantages of a tilt-rotor were evident as early as the 1930s and 1940s, when British and German inventors outlined the vehicle's basic configuration. The idea was picked up in the United States by the Platt-LePage Company in the 1940s and actually put into operation by the Transcendental Aircraft Corporation, established by Mario A. Guerrieri and Robert L. Lichten. Their vehicle, the prototype

85. Spenser, *Whirlybirds*, pp. 252–273; John J. Tolson, *Airmobility 1961–1971*, Vietnam Studies (Washington, DC: Department of the Army, 1989); first printed in 1973 as CMH Pub 90-4. The adoption of helicopters by the Army for tactical operations was foreseen well before the Vietnam War: see General James M. Gavin, "Cavalry, and I Don't Mean Horses," *Harpers* (April 1954): 54–60.

Model 1G, flew test flights in 1954 and 1955 and was the first working example of a tilt-rotor.[86]

The Bell Aircraft Company, which since World War II had pursued leading-edge aeronautical technologies (such as high-speed experimental aircraft and helicopters), hired Robert Lichten to head up research on tilt-rotors. Not long thereafter, the U.S. Army and U.S. Air Force sought proposals for a convertiplane (a V/STOL aircraft). Bell submitted Lichten's tilt-rotor design and won a contract for two prototypes named the Bell XV-3. The first flight of the XV-3 took place in 1955, and tests quickly highlighted problems with the vehicle's dynamic stability. In flight, the XV-3 would begin shaking; during a 1956 test, the vibrations were so severe that the pilot blacked out. Searching for the cause of the shaking, Bell, the Army, and the Air Force took the XV-3 to Ames for testing. Thus began the NACA's, and then NASA's, long association with tilt-rotor technology.[87]

The particular arrangement of the tilt-rotor, it turns out, makes the vehicle susceptible to what is called aeroelastic instability. In simpler terms, small vibrations in the rotors can propagate through the engine, through the wingtip pylon, and on up the wing to the fuselage. All vehicles experience vibrations as loads vary, but in stable vehicles, these vibrations decrease as the energy travels through the structure. With the XV-3, the vibrations could grow to dangerous proportions. Beginning in 1957, Ames engineers began a year and a half of testing that included running the XV-3 in the large 40-by-80-foot wind tunnel. After making a number of modifications, many of them increasing the stiffness of the rotor-pylon-wing assembly, flight testing resumed in 1958 with the XV-3 finally achieving full conversion from helicopter mode to airplane mode. After Bell delivered the tilt-rotor to the Air Force, and after the Air Force conducted its own tests at Edwards Air Force Base, the XV-3 returned to Ames, where engineers could more closely examine its performance and complex aerodynamics. In 1968, Bell conducted another round of tests of the XV-3 in the Ames 40-by-80 to verify aeroelastic modeling (i.e., to see how actual vibrations agreed with predictions from Bell's models). The tests went well until the XV-3 suffered a catastrophic structural failure, one linked in post-test analysis to metal fatigue.[88]

86. Maisel et al., *The History of the XV-15 Tilt Rotor Research Aircraft*, pp. 6–11.

87. Ibid., pp. 4–5, 11–14.

88. Ibid., pp. 4–5, 11–16.

In 1969, NASA and the U.S. Army agreed to jointly fund and operate two rotorcraft laboratories, one at Ames and one at Langley.[89] In 1971, Ames created a V/STOL Project Office. This reflected the accumulation of past experience with V/STOL research, including some of the aforementioned prototype vehicles, and new work such as the augmentor wing.[90] In 1971, the Army and NASA agreed to build a tilt-rotor proof-of-concept vehicle. Headquarters agreed to the tilt-rotor project only on the grounds that it receive Army backing and joint funding.[91] The joint funding, the authors of the official tilt-rotor history argue, kept the project going in part because neither party wanted to be the first one to back out, though by 1978, in the face of growing costs, only timely funding from a third party, the Navy, kept the program rolling.[92] In 1972, Ames completed the institutional foundation for the tilt-rotor with a Tilt Rotor Research Aircraft Project Office. Indicative of the level of organizational control that had previously been the hallmark of NASA's space projects, Headquarters now required "a Project Development Plan, a Risk Assessment, an Environmental Impact Statement, a Safety Plan, a Reliability and Quality Assurance (R&QA) Plan, and a Procurement Plan."[93]

In 1972, NASA Ames gave contracts to Boeing and Bell for design studies of a new prototype tilt-rotor incorporating everything that had been learned to date. Grumman and Sikorsky also bid on the project. Boeing's design was hingeless, meaning that it tilted only the proprotor, not the engine. Bell's design rotated the proprotor and the engines. Boeing's design cut costs by making use of a Mitsubishi MU-2J fuselage. Bell won the competition and was given a contract for two XV-15 vehicles with a target cost of $26.4 million.[94]

For the next five years, Bell tackled the design and fabrication of the prototypes and, as with previous research vehicles, worked with the researchers at Ames. Rockwell served as a subcontractor to Bell, building the fuselage and

---

89. Ibid., p. 19; Victor L. Peterson, interview by Robert G. Ferguson, tape recording, Los Altos, CA, 17 January 2005, NASA HRC.
90. Maisel et al., *The History of the XV-15 Tilt Rotor Research Aircraft,* p. 28.
91. Ibid., pp. 30–31.
92. Ibid., p. 59. Victor Peterson, speaking in general, said that the joint nature of the NASA-Army lab made it a stronger entity in the Agency's competition for resources. Peterson interview.
93. Maisel et al., *The History of the XV-15 Tilt Rotor Research Aircraft,* p. 32.
94. Ibid., pp. 27, 35, 37. George Low noted that "[t]he selection was a fairly straightforward one, with Bell rated considerably higher on mission suitability, and with Bell's most probable cost being less than Boeing[']s. However, the actual bid price of Boeing was slightly lower than Bell's." George M. Low to James C. Fletcher, "Activities During Week of April 8–14, 1973," 17 April 1973, Fletcher Correspondence, folder 4248, NASA HRC.

(NASA image ECN-13850)

Figure 4.3. The XV-15 tilt-rotor aloft for its first flight at Dryden in 1980.

empennage.[95] The transmission proved to be one of the more difficult aspects of the vehicle. Not only did the transmission need to reduce the revolutions per minute, or rpm, for the proprotors, but the system was cross-linked with the opposite proprotor in case of an engine failure (the vehicle was designed to be highly fault tolerant). The transmission required many design and manufacturing changes for the sake of reliability.[96]

In May 1977, the XV-15 flew for the first time.[97] Bell conducted the early flight testing at its test center at Arlington Municipal Airport in Texas. After wind tunnel tests, Bell conducted another round of flight tests that saw the full conversion to aircraft mode in July 1979. Testing moved to Dryden in late 1980, where NASA accepted the vehicle from Bell and continued testing for the next year.[98] Ames monitored all the flight tests, including those at Dryden.

95. Maisel et al., *The History of the XV-15 Tilt Rotor Research Aircraft,* p. 49.

96. Ibid., pp. 43–44, 47.

97. Ibid., pp. 55–56.

98. Ibid., pp. 59–63.

The Project Office contracted with a Sunnyvale firm to create a digital test database that allowed engineers to sift through and query data. Ames made use of this database for other programs such as Black Hawk helicopter tests and the QSRA Jump-Strut Project.[99]

Since the tilt-rotor was so unconventional, NASA was testing some flight characteristics for the first time. For example, what would happen to the proprotors' downwash when it reacted with the rest of the vehicle or when the XV-15 was in a hover? NASA also put closure on the phenomenon that had brought the XV-3 to Ames in the first place, aeroelasticity. To this end, engineers installed "excitation" actuators to induce oscillation and confirm that the vehicle properly damped the vibrations.[100]

After NASA and the Army concluded their initial tests, they returned one of the two XV-15 vehicles to Bell in 1981 under a bailment agreement (the government retained ownership while Bell was responsible for the vehicle and its operations).[101] Meanwhile, NASA contracted with Boeing to design improved rotors that took advantage of better materials and offered higher performance. Ground testing and flight tests took place in 1987. In 1992, Bell crashed the vehicle it had under bailment and so took over the remaining XV-15 under a similar agreement.[102] By the conclusion of the entire tilt-rotor research program in 1993, costs came to $50.4 million, with Bell investing $1.5 million.[103] Finally, after decades of research on the tilt-rotor configuration, Bell and Boeing began manufacturing the V-22 Osprey for the Marine Corps and Air Force, with the test articles flying in 1989 and the first operational deployment taking place in Iraq in 2007.

STOL

STOL aircraft are conventional aircraft with unconventional lift capabilities. Unlike V/STOL aircraft, which can take off and land vertically (but use STOL to save fuel and increase payload), STOL aircraft need some amount of runway. Most runways are in the range of 8,000 feet long; a STOL aircraft can safely use a runway a fraction of that length down to about 1,000 feet long.[104] The

99. Ibid., p. 52.
100. Ibid., pp. 67–70.
101. Ibid., p. 90.
102. Ibid., pp. 77–82, 98–99.
103. Ibid., pp. 103–105.
104. W. H. Deckert and J. A. Franklin, *Powered-Lift Aircraft Technology* (Washington, DC: NASA SP-501, 1989), p. 3. It should be noted that STOL-equipped bush aircraft routinely use even shorter runways.

technologies used to achieve STOL range from high-lift flaps and slats (devices the Centers had been studying since the NACA days) to power-lift arrangements in which propellers or jet engines directly contribute to the production of lift as opposed to merely providing forward thrust.

NASA's STOL research entered into a highly active and visible phase in the 1970s. Some technical advances did make STOL more attractive, but this research phase coincided with a period of heightened interest in commercial STOL. Military aircraft have long had immediate use for STOL technology, and while the NASA program was principally motivated by the military, it began to envision and sell the research as a potential solution to crowded airports and flight routes, especially in the northeast corridor. Although the technology was hardly guaranteed, it attracted its share of proponents, mainly from the FAA and the airlines. In 1968, Joan Barriage, the FAA Program Manager for VTOL and STOL, made this unfortunate prophecy:

> Some 40 years from now, historians charged with planning a Wright Centennial may well find themselves hard pressed to identify a more significant aeronautical breakthrough in 20th Century U.S. air transportation than instituting a STOL and VTOL system.[105]

The STOL and VTOL air transportation system never happened. The irony, insofar as NASA is concerned, is that despite a successful research program, STOL's commercial application hinged on the creation of a new air system. Without STOLports, STOL aircraft would be operating from the same crowded airports as conventional aircraft, and they would do so at a higher cost. Creating a STOL system was no small matter, even with the large number of underutilized small airstrips that dot the country. A STOL system would have required new flight patterns and STOL-configured instrument landing systems. Most importantly, STOL aircraft would have had to overcome much of the same community opposition that had deterred the building of new conventional airports in the first place. Even if they could operate from smaller strips, they were still as loud as, and potentially louder than, conventional aircraft.[106]

105. Joan B. Barriage, "STOL and VTOL Air Transportation—From the Ground Up," *Astronautics and Aeronautics* (September 1968): 44–52.

106. In testimony given to the House, NASA's Acting Director for STOL Technology noted that there were about 400 airports within a 100-mile radius of New York City that could operate as STOL ports. Gerald G. Kayten, Acting Director, STOL Technology Office, OART, NASA, before

Oscar Bakke, the FAA Director of the Eastern Region, was an influential STOL proponent. He argued for STOL to serve New York City and organized, in 1966, a two-day exercise to supply the city using aircraft landing in small areas, including parks and waterfront piers. Part of the rationale for the exercise was to "demonstrate the capability of V/STOL…aircraft to provide air access…," but it was also meant to spur municipal planners into building V/STOL facilities. Ben Darden, also of the FAA, followed in Bakke's footsteps and advocated a STOLport on top of a New York building located next to the water. The City of New York went so far as to issue a request for a proposal (RFP) to study the idea in 1969.[107]

In late 1967, the Civil Aeronautics Board called for an investigation into the feasibility of establishing a northeast corridor V/STOL system.[108] In 1968, the FAA began its own series of tests at its National Aviation Facilities Experimental Test Center (NAFEC) in Atlantic City, New Jersey, to examine the performance of STOL aircraft and develop federal standards for their approval and operation. The FAA tests showed that STOL aircraft could achieve a 10.5-degree glide slope and that 6 to 9 degrees was practical (two to three times the normal glide slope).[109] In the same year, LaGuardia airport opened a STOL runway. A mere 1,095 feet long, it did not have any instrument approach and so required visual flight rule conditions (i.e., clear skies).[110]

Eastern Airlines took advantage of the LaGuardia STOL runway and used a French-made Breguet 941 to study, without passengers, the possibility of

---

the Committee on Science and Astronautics, House of Representatives, March 1971, "STOL Technology," folder 11726, NASA HRC. See also Barriage, "STOL and VTOL Air Transportation—From the Ground Up," pp. 44–52. Robert W. Rummel, TWA vice president, projected in 1970 that first-generation STOL aircraft would have higher direct operating costs than a large supersonic jet. See "Applications Decision Is Key Hurdle for V/STOL," *Aviation Week & Space Technology* (22 June 1970): 144–150.

107. Federal Aviation Administration, Department of Transportation, "A Place to Land," undated pamphlet describing November 1966 exercise in New York City, STOL, folder 11726, NASA HRC; Bruce H. Frisch, "Why New York Has No Stolport," *Aeronautics and Astronautics* (December 1970): 22–23; H. Watts Bagley, "STOL—Aviation's Sleeping (?) Beauty," *Astronautics and Aeronautics* (September 1968): 36–38.

108. Bruce Frisch, "Getting V/STOL Services Going," *Astronautics and Aeronautics* (September 1968): 40–43.

109. Bagley, "STOL—Aviation's Sleeping (?) Beauty," pp. 36–38.

110. Wilson Leach, "STOL Runway Opens at LaGuardia," *Business & Commercial Aviation* (October 1968): 122–123.

a New York–Washington, DC, STOL route.[111] Eastern did not implement commercial STOL service, but Scott Crossfield, the former NACA/NASA test pilot and subsequent vice president for research and development at Eastern Airlines, became an advocate for a government-sponsored STOL system. From Eastern's perspective, he wrote in 1970, it would allow the airline to avoid the costs of air traffic congestion (which Eastern estimated at $1 million a week in nonproductive flying).[112] Eastern was not alone in its desire for greater infrastructure investment. Najeeb E. Halaby, Pan American Airways' president (after he left his post at the FAA), urged the creation of a STOLport built on piers over the Hudson River.[113]

NASA researchers echoed many of the same concerns about aviation traffic and the potential for STOL, though NASA's comments tended to be more circumspect about the actual technical challenges.[114] Bradford Wick of Ames and Richard Kuhn of Langley wrote in 1971 that although the original goal of V/STOL technology was to operate from downtown locations, probably using VTOL, STOL was a more likely development because of recent increases in airport congestion. STOLports would take much less space and thus "be easier to finance and develop" than traditional airports.[115] *Aviation Week & Space Technology*, reporting on an Ames STOL conference in 1972, noted that Ames's Dr. Leonard Roberts argued that commercial airport traffic would come to a head by the end of the 1970s and that STOL aircraft could help reduce the problem.[116]

NASA's actual research on STOL technology long predated the late-1960s-to-early-1970s enthusiasm for commercial STOL. Langley researched externally blown flaps, one of the few STOL technologies that has seen practical application, as early as the 1950s.[117] From 1959 to 1967, Ames researchers

111. "Can a STOL Go Commercial?" *Business Week* (7 September 1968).
112. A. Scott Crossfield, "Short-Haul STOL Concepts in Perspective," *Astronautics and Aeronautics* (December 1970): 44–48.
113. Bagley, "STOL—Aviation's Sleeping (?) Beauty," pp. 36–38.
114. Compare, for example, NASA's assessment of STOL with other promotional articles in the September 1968 issue of *Astronautics and Aeronautics*; NASA Langley Research Center, "VTOL and STOL Technology in Review," *Astronautics and Aeronautics* (September 1968): 56–67.
115. Bradford H. Wick and Richard E. Kuhn, "Turbofan STOL Research at NASA," *Astronautics and Aeronautics* (May 1971): 32–50.
116. Richard G. O'Lone, "Noise Main Challenge in STOL Research," *Aviation Week & Space Technology* (30 October 1972): 16–17.
117. John P. Campbell and Joseph L. Johnson, Jr., "Wind-Tunnel Investigation of an External-Flow Jet Augmented Slotted Flap Suitable for Application to Airplanes with Pod-Mounted Jet Engines," NACA Technical Note 3898, 1956.

(NASA image AC73-2101)

Figure 4.4. A de Havilland Buffalo modified with an augmentor wing and turbofans performs a short takeoff in 1973.

performed STOL experiments using boundary layer control on two aircraft: a YC-134A and an NC-130B.[118] In 1971, Ames began testing a North American Rockwell OV-10A with a rotating-cylinder flap. This device was, literally, a long, rotating cylinder just forward of the wing flap and aileron. Hydraulically driven, the cylinder maintained the boundary layer over the wing, thus allowing for lower maneuvering speeds and reduced runway distances.[119]

The most influential STOL work, however, came from two projects, the augmentor wing and the Quiet Short-Haul Research Aircraft (QSRA). The first project grew out of a 1965 collaboration with de Havilland, a Canadian firm, and the Defence Research Board of Canada. The Canadians had been researching a concept called the augmentor wing since 1961. An augmentor wing takes jet engine bleed air and channels it through ducts to the wing's trailing edge and flaps. Combined with the action of blowing air over the wing's flaps, the system increases lift and provides a margin of safety in case of an engine failure

118. Paul F. Borchers, James A. Franklin, and Jay W. Fletcher, *Flight Research at Ames: Fifty-Seven Years of Development and Validation of Aeronautical Technology* (Washington, DC: NASA, 1998), pp. 48–49.

119. Ibid., p. 50.

through cross-ducting the bleed air. Following joint studies using NASA's wind tunnels, NASA renewed the joint venture in 1970, this time with the Canadian Department of Industry, Trade, and Commerce (CDITC). NASA and the CDITC retrofitted a de Havilland Buffalo aircraft with two Rolls-Royce Spey turbofans.[120] The Augmentor Wing Jet STOL Research Aircraft (AWJSRA), as it was named, began test flights in 1972. The AWJSRA was the world's first jet STOL aircraft, managing takeoffs and landings in less than 1,000 feet of runway. After validating the augmentor wing and the aircraft's controllability, Ames engineers went on to add digital guidance controls and automatic approach and landing equipment.[121]

Following the AWJSRA, Ames sought to test a new STOL design that would be more efficient and quiet. As early as 1968, Langley staffers had identified noise as a new, key issue in the development of commercial STOL. They wrote, "Noise certainly presents one of the more critical problems for the VTOL and STOL aircraft. Up to this point in their development, the primary emphasis has been on performance, with little or no attention to achieving acceptable noise levels.... [T]here will need to be trade-offs between performance and noise."[122] The Center's first run at funding for this line of work envisioned a combined quiet STOL (QUESTOL) and quiet jet (QCSEE) program. Funding for this, however, was canceled in 1972 as a cost-cutting measure because the White House believed that NASA could do the same research on two STOL aircraft that were already under way in the Air Force's Advanced Medium STOL Transport (AMST) program. Boeing received a contract to build the YC-14 and McDonnell Douglas a contract for the YC-15. Although both of these aircraft eventually flew, neither was designed for "quiet" commercial operation.[123]

---

120. Wick and Kuhn, "Turbofan STOL Research at NASA," pp. 32–50; Deckert and Franklin, *Powered-Lift Aircraft Technology*, p. 4. The Canadians actually held the patent on the augmentor wing; see William S. Aiken, Jr., to RD-T/Deputy Associate Administrator—Technology, "Extension of Joint Agreement with Canada on Augmentor Wing Research," 31 March 1972, STOL, folder 11736, NASA HRC.
121. Borchers et al., *Flight Research at Ames*, p. 51. On noise, "Community acceptance is essential to viability of an operational STOL system. The quiet propulsion retrofit will provide the test bed for public acceptance and for establishment of noise certification standards." Gerald G. Kayten, Acting Director, STOL Technology Office, OART, NASA, before the Committee on Science and Astronautics, House of Representatives, March 1971, "STOL Technology," folder 11726, NASA HRC.
122. NASA Langley Research Center, "VTOL and STOL Technology in Review," pp. 56–67.
123. Representative Ken Hechler et al., telegram to Vice President Spiro Agnew, 13 December 1972, STOL, folder 11726, NASA HRC; James C. Fletcher to John L. McLucas, 20 September

Despite this setback, Ames eventually received funding for a more modest quiet STOL program called the Quiet Short-haul Research Aircraft (QSRA). For this design, Ames contracted with Boeing to build a new wing for the de Havilland Buffalo. Instead of channeling the air internally, the QSRA's four engines were mounted to blow over the top of the wing (i.e., "upper surface blowing"). Flight testing began in 1978. The QSRA had very good performance, with a perceived noise level lower than for comparable conventional aircraft. In 1980, with Navy support, pilots performed takeoffs and landings on the deck of the aircraft carrier USS Kitty Hawk.[124]

As noted, the greatest obstacle for commercial STOL aircraft was not the technology, but the infrastructure. Without a comprehensive STOL system in place, STOL jets made little economic sense for the airlines. One company, however, did pursue the technology: de Havilland Aircraft of Canada. The firm, which would eventually become part of snowmobile maker Bombardier, built STOL technology into its Dash-7 aircraft, a four-engine turboprop produced in low numbers from the 1970s to the late 1980s. Instead of flying into STOLports and reducing aviation congestion, the Dash-7 was most at home in rugged flying conditions such as small, high-altitude airstrips.[125] Powered lift technology saw application in the McDonnell Douglas C-17 transport (now the Boeing C-17). Here again, this was an aircraft designed for rugged airstrips rather than crowded urban corridors.[126]

NASA STOL research was, ultimately, successful insofar as the programs achieved their stated goals. The technology was appropriate and readily adapted to military aircraft and niche vehicles traveling into rugged airstrips. Commercial STOL was another matter. Expectations about commercial STOL certainly helped enable programs like the QSRA, but it is also true that NASA, especially Ames, was committed to STOL as a Center specialty, regardless of end use. Ultimately, one of the factors that weighed against commercial STOL (beyond the major obstacle of adding infrastructure) was that aviation gridlock did not occur, at least not in the 1970s.[127]

1973, STOL, folder 11726, NASA HRC; James C. Fletcher to John L. McLucas, 2 August 1973, STOL, folder 11726, NASA HRC.

124. Borchers et al., *Flight Research at Ames*, p. 52; Muenger, *Searching the Horizon*, pp. 269–271.

125. Ken Botwright, "Stalled STOL Is Taking Off in Canadian Turbo-Prop Job," *Washington Post*, 21 January 1973, STOL, folder 11726, NASA HRC.

126. Deckert and Franklin, *Powered-Lift Aircraft Technology*, pp. 1–3.

127. There are numerous explanations for why aviation congestion has not resulted in the predicted gridlock. Among the shifts that have occurred since the early 1970s are airline

The Rise and Consolidation of Rotorcraft Research

The year 1969 saw the creation of joint Army-NASA rotorcraft laboratories at both Ames and Langley. These laboratories were never large by NASA standards, but the Army's participation lent itself to the creation of a stable, long-term community of rotorcraft-focused scientists and engineers. The Army's demand for a strong R&D base for helicopters, an outgrowth of the rise of helicopters in Army operations, gave scientists and engineers a well-defined research scope. Interestingly, the unique problems faced by rotorcraft researchers also gave this work a distinct, insular character.[128]

Two physical factors made, and continue to make, rotorcraft very difficult to model. The first problem is that rotor blades experience vast changes in lift as the helicopter moves forward. The advancing blade is traveling at the sum of the blade's rotational velocity plus the forward velocity of the helicopter. The retreating blade is traveling at the sum of the forward velocity of the helicopter less the rotational velocity of the blade. In one rotation, a blade can approach the limits of compressibility on one side and then stall on the other side. The second problem is that rotor blades move through their own wake. To model this requires knowing how turbulence propagates long after a blade has passed. For both these reasons, rotorcraft research has maintained a strong reliance on wind tunnel testing, especially large and full-scale testing.

With access to the massive 40-by-80-foot wind tunnel, the Army Aeronautical Research Laboratory (as it was initially named) at Ames was well equipped. The fact that the large tunnel operated at low subsonic speeds did nothing to hinder rotorcraft work; helicopters travel at low subsonic speeds anyway. The NACA first used the tunnel for rotorcraft research in 1953. In addition to the 40-by-80, Ames made use of the 7-by-10-foot subsonic tunnel.[129] At Langley, rotorcraft researchers worked closely with the Transonic Dynamics Tunnel (TDT), a tunnel that dated from 1938 but had been updated in the 1950s with a slotted throat to allow research in the transonic regime. The TDT's first helicopter research was in 1963. One of Langley's areas of expertise was aeroelasticity and aerostructures. To this end, the TDT came to be an important tool in understanding rotorcraft

---

deregulation, which resulted in dramatically altered route structures; changes in airline load factors; and continuing optimization of the FAA's air traffic control.

128. Wayne Johnson, telephone interview by Robert G. Ferguson, 30 May 2007, NASA HRC; William Warmbrodt, telephone interview by Robert G. Ferguson, 14 May 2007, NASA HRC; Chee Tung, telephone interview by Robert G. Ferguson, 20 April 2007, NASA HRC.

129. William Warmbrodt, Charles A. Smith, and Wayne Johnson, "Rotorcraft Research Testing in the National Full-Scale Aerodynamics Complex at NASA Ames Research Center" (NASA TM-86687, May 1985).

aeroelasticity (that is, the flutter and loads in rotor systems).[130] In addition to the aeroelastic challenges of tilt-rotors, new technologies such as hingeless rotors and bearingless rotors required that researchers develop more comprehensive methodologies and models for aeroelasticity.[131]

NASA consolidated the Ames and Langley rotorcraft activities at Ames in 1978. This was certainly a blow to Langley as the Center had a long and rich tradition of rotorcraft work.[132] All of the Langley flight-test vehicles went to Ames, as did a small number of researchers. Interestingly, Langley continued to perform rotorcraft research; the Army's lab did not close, and specialized tunnels like the TDT, as well as the 14-by-22-foot tunnel, continued to do essential rotorcraft testing on behalf of the Army and the Ames group. The decision to consolidate at Ames was in line with the division of labor that had been forming during the decade.[133]

Terminal Configured Vehicle

The Terminal Configured Vehicle (TCV) Program was Langley's marquee program for conducting research on avionics and flight procedures in commercial aircraft. The program coalesced in the early 1970s and reflected a belief that the time was ripe for exploiting emerging electronic and digital instruments in the cockpit. Researchers argued that new avionics and routines could ease airport congestion, make commercial aircraft safer, and reduce noise. Another timely selling point was that reducing terminal area congestion naturally led to greater fuel efficiency. To do all of this in a meaningful way, Langley's researchers sought a flying laboratory, one in which they could test whole cockpit systems safely, all the while monitoring the performance of the pilots. For the TCV, Langley purchased Boeing's first 737, a test aircraft that would have been too expensive for Boeing to overhaul for commercial service. Boeing, instead,

130. William T. Yeager, Jr., and Raymond G. Kvaternik, "A Historical Overview of Aeroelasticity Branch and Transonic Dynamics Tunnel Contributions to Rotorcraft Technology and Development" (NASA TM-2001-211054, ARL-TR-2564, August 2001).

131. Robert A. Ormiston, William G. Warmbrodt, Dewey H. Hodges, and David A. Peters, "Survey of Army/NASA Rotorcraft Aeroelastic Stability Research" (NASA TM-101026, October 1988).

132. A summary of many of the major rotorcraft programs can be found in W. J. Snyder, "Rotorcraft Flight Research with Emphasis on Rotor Systems," in *NASA/Army Rotorcraft Technology*, vol. 3, *Systems Integration, Research Aircraft, and Industry*, NASA Conference Publication 2495, pp. 1234–1273.

133. Wayne Johnson interview.

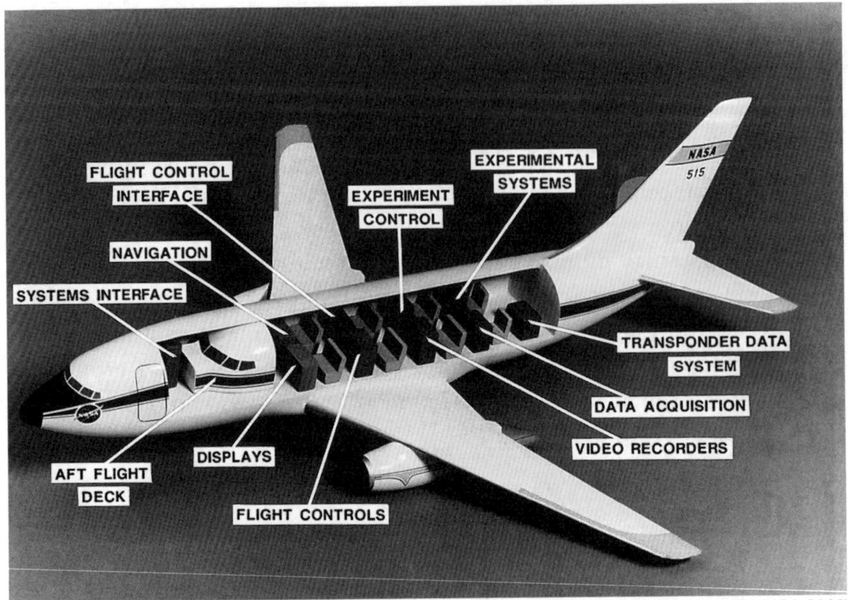

(NASA image EL-1996-0082)

Figure 4.5. Cutaway of Langley's Terminal Configured Vehicle showing the second cockpit and research equipment.

renovated the aircraft for the TCV program and, through the 1970s, assisted with NASA's avionics experiments.[134]

The name "Terminal Configured Vehicle" came about because the problems it sought to address were with the terminal area (landing) with a special concern for safe and efficient operation during periods of congestion and poor weather. An early program document noted that terminal operations accounted for "more than half of all fatal accidents." Langley's program was careful to emphasize that it complemented FAA and Department of Transportation research into new air traffic control systems and procedures. NASA's purview was the "airborne portion of the system."[135]

Part of the impetus for acquisition of the 737 was that Boeing had been working on advanced avionics for the SST when Congress canceled the

134. "Program Plan for Terminal Configured Vehicle" (NASA TM-1082227, 1 December 1973); Lane E. Wallace, *Airborne Trailblazer: Two Decades with NASA Langley's 737 Flying Laboratory* (Washington, DC: NASA SP-4216, 1994).

135. "Program Plan for Terminal Configured Vehicle" (NASA TM-1082227, 1 December 1973). NASA and the DOT/FAA signed a memorandum regarding their respective roles and cooperative efforts.

program in 1971. Langley's plans to deploy a flying avionics laboratory gave Boeing an opportunity to cooperate with NASA on the flight testing of the SST's digital electronic displays. When Boeing renovated the 737, it created a second, aft cockpit that contained the new avionics. It was wired such that control could be transferred between the cockpits, while researchers and test equipment occupied the remaining passenger space.[136] The TCV became part of a much larger development and test network that comprised ground facilities at Langley and a flight-test range at Wallops Island. Among the ground facilities used by the TCV program were a TCV simulator (a copy of the TCV's aft cockpit), a visual motion simulator, a terminal area air traffic model for simulating air traffic control operating environments, and the Experimental Avionics System Integration Laboratory. Wallops had a specially equipped airfield with an acoustic noise range as well as a laser and radar tracking system. Together, the Langley and Wallops facilities allowed for development and testing to begin on the ground, flight testing to take place at Wallops, and then data analysis to occur back at Langley.[137]

At the outset, the broad program goals of the TCV were to develop systems and procedures for zero-visibility landings, increased airport throughput (i.e., more landings per hour), reduced cockpit workload, and reduced environmental impact.[138] What this came down to in terms of specific investigations was a wide range of technologies and operating procedures, especially those concerned with the display of information in the cockpit. The TCV's aft cockpit came equipped with two new and important instruments: the electronic attitude director indicator (EADI) and an electronic horizontal situation indicator (EHSI). The former was much like an electronic version of a typical attitude indicator (i.e., an artificial horizon), while the latter was much like a moving map display. While these would become commonplace in commercial aircraft beginning in the 1980s, the TCV's versions were special because they could be customized to show additional information or to portray it in a different manner. Two examples serve to show the versatility of the system.

With traditional avionics, a pilot performs an instrument landing by watching two needles on the artificial horizon: one is the glide slope, the other the localizer. As the needles move up or down, or left or right, the pilot makes adjustments in order to keep the aircraft on a proper descent to the runway. These same indicators were represented on the TCV's EADI, but Langley's

136. Wallace, *Airborne Trailblazer*, pp. 12–13.

137. "Terminal Configured Vehicle Program: Test Facilities Guide" (Langley Research Center: NASA SP-435, January 1980).

138. "Program Plan for Terminal Configured Vehicle" (NASA TM-1082227, 1 December 1973).

researchers thought there might be a better way. They chose to superimpose a simple graphical image of a runway, in perspective, on the EADI. The image would move left or right, for example, depending on whether the aircraft was to the right or left of the runway centerline. This all seems eminently simple now, but it was groundbreaking at the time. The pilots reported that having the runway integrated onto the display gave them a better understanding of their position, and they were able to track the localizer beacon consistently.[139] A second example is the use of the EADI to display traffic information. Langley conducted this work in conjunction with the FAA near the end of the decade. Researchers superimposed a map of simulated traffic around the aircraft on the EADI while pilots flew approaches and responded to traffic as it neared them. Researchers sought to learn what information the pilots needed most and how to encode this on the map. For example, researchers found that pilots were most interested in altitude coding (i.e., symbols showing whether the traffic was above or below).[140] While the TCV conducted a wide range of studies, most were very similar to these examples insofar as they made use of the aircraft's highly flexible digital displays and sought to evaluate both pilot performance and opinion. As much as NASA attempted to delineate its own work by focusing on the airborne part, much of the research necessarily involved coordination with the FAA.

The TCV, eventually renamed the Transport Systems Research Vehicle, continued serving Langley into the 1990s. It performed testing for the FAA's microwave landing system. It tested global positioning navigation systems. It explored the dangers of wind shear. When the aircraft was finally nearing the end of its useful life, NASA cast about and purchased a Boeing 757 in 1994, calling it the Airborne Research Integrated Experiments Station (ARIES). The ARIES program anticipated building an aft cockpit similar to that on the 737, but budget constrains prevented this.[141]

---

139. S. A. Morello, C. E. Knox, and G. G. Steinmetz, "Flight-Test Evaluation of Two Electronic Display Formats for Approach to Landing Under Instrument Conditions" (Langley Research Center: NASA TP-1085, 1 December 1977).
140. T. S. Abbot, G. C. Moen, L. H. Person, Jr., G. L. Keyser, Jr., K. R. Yenni, and J. F. Garren, Jr., "Early Flight Test Experience with Cockpit Displayed Traffic Information (CDTI)" (Langley Research Center: NASA TM-80221, 1 February 1980).
141. Michael S. Wusk, "ARIES: NASA Langley's Airborne Research Facility" (AIAA Aircraft Technology Integration and Operations Forum, Los Angeles, CA, 1–3 October 2002).

Conclusion

The 1970s proved highly productive for NASA's aeronautics research community. Arguably, much of this was the result of seeds sown in earlier decades. One may choose to draw attention to the continuities in research—to point out, for example, the slow evolution of computational capability that eventually became a full-blown computational fluid dynamics program—but the inconsistencies are there as well. NASA's aeronautics in the 1970s appears more responsive to societal needs and more programmatically organized. The original distinctions made in 1958 between mission-oriented Centers and research-oriented Centers blurred in the 1970s. Even the research-oriented Centers were positioning themselves to address specific areas of technological application, seeking support for packaged research programs and creating centers of excellence meant to generate constant levels of support (regardless of the specific yearly research activity). Taking one step further back, aeronautics weathered the post-Apollo malaise. Though the Agency suffered very large funding cutbacks early in the decade, the impact was primarily on the space program. For the most part, the difficult shift from Apollo to the Space Shuttle did not lead to similar discontinuities in aeronautics.

Chapter 5:

Cold War Revival and Ideological Muddle

Like most institutions, NASA has had an organizational momentum that resisted political and social change. As its experience in the 1970s with energy efficiency showed, the Agency grew adept at bridging the political and the scientific sides of its house. President Ronald Reagan, however, ushered in a new and paradoxical era for federally funded civilian research. On the one hand, Reagan inveighed against "big government," and at his back was an emboldened group of free-market-oriented economists who were ideologically opposed to federally funded civilian research. On the other hand, Reagan's two terms brought greatly increased levels of funding for new weapons systems and, in his second term, specific funding for a large-scale civilian aeronautics project, the National Aero-Space Plane (NASP). Over the short term, the ideological objections took a back seat to an invigorated Cold War funded by deficit spending. The pipeline of new aircraft, some of them designed and built in complete secrecy (i.e., "black" programs) helped keep NASA busy through the decade. Over the long term, the ideological objections to federally funded civilian research would only grow stronger.

While aeronautics received a fraction of NASA's entire budget, it was among the most vulnerable to the Agency's ideological critics. Unlike the human space program, which had no market counterpart, aeronautics research was deeply tied to private industry.[1] One of the ironies of the Reagan-era intention to curtail the applied side of aeronautical R&D and focus on basic research and military R&D is that only a decade earlier, NASA had bent over backwards to show that it was relevant and applied. The 1970s-era programs that emphasized technology transfer and directly funded industrial R&D programs (such as parts of the ACEE program) invited concern over corporate subsidy and market

1. The absence of commercial investment in space (beyond communications satellites) led to the White House's 1984 National Policy on the Commercial Use of Space and the creation of the Office of Commercial Programs at NASA. Rumerman, *NASA Historical Data Book*, vol. 6, p. 355.

interference. The Agency was being whipsawed. More troubling from the perspective of the researchers at the Centers was the battle over aeronautics being waged among the Office of Management and Budget, the Office of Science and Technology Policy, Congress, and influential groups such as the National Research Council's Aeronautics and Space Engineering Board (ASEB). This was a far cry from the days in which the Centers established their research direction and then sought budgetary approval from Headquarters. Now, even Headquarters appeared to be losing its seat at the table.

The administrative distance between Headquarters and the Centers also grew through the decade. Whereas Centers had reported directly to the NASA Administrator, starting in 1981 they reported to the Associate Administrator of the appropriate program office, which in the case of aeronautics was the Office of Aeronautics and Space Technology (OAST).[2] This administrative distance is important when one considers the role that Center Directors traditionally had as advocates for Center-derived research programs. The picture becomes even more complicated through the 1980s: even as the Centers become further removed from the Administrator and Associate Administrator, there is a multiplication of other offices at Headquarters, each one diverting the Administrator's attention.

In 1984, OAST itself underwent reorganization. Since 1972, OAST had had three primary divisions: Aerospace (which oversaw basic research programs), Aeronautical Systems (which oversaw specific vehicle and system programs), and Space Systems (which, similarly, oversaw specific space vehicles and systems). There were also managerial and institutional divisions, an energy systems division (eliminated in 1983), and the individual Center hierarchies. With the 1984 reorganization, the Aerospace Division saw its disciplines elevated and divided into their own divisions: Aerodynamics; Information Sciences and Human Factors; Materials and Structures; Propulsion, Power and Energy; and Flight Projects. Vehicle and system programs remained under either the Aeronautics or Space Directorate.[3]

At the Centers, the disciplinary jostling that had been taking place over the previous two decades between Ames and Langley had finally reached some measure of stability. Ames had solidified its positions in supercomputing and the related area of flight simulation, insofar as it was highly computational. Ames had also acquired the short-haul and rotorcraft programs and had established diversified, non-aeronautics programs in the astrophysical and biological science programs. Langley's focus, meanwhile, contracted to long-haul fixed-wing

2. Ibid., pp. 388–389.

3. Ibid., pp. 185–188.

aircraft and expertise in aerodynamics, materials and structures, guidance and control, and environmental quality.[4] The Dryden Flight Research Center lost its status as a full-fledged Center in 1981 when it was placed underneath Ames, while Ames was asked to transfer most of its own flight-test activities to Dryden. Renamed the Ames-Dryden Flight Research Facility, Dryden would eventually regain its status (and old name) in 1994. This was part of a larger effort to streamline and consolidate NASA facilities. NASA Goddard similarly took over the Wallops Flight Facility.[5]

The Centers undertook a number of important renovations of test facilities during the 1980s. Many of the wind tunnels built in the 1940s and 1950s were becoming quite dated. At Ames, the 12-Foot Pressure Wind Tunnel as well as the 40-by-80-Foot Wind Tunnel underwent renovations. The latter added an open circuit section that increased the size of the tunnel to 80 by 120 feet. Lewis renovated its Unitary Plan Tunnel, while Langley opened its National Transonic Facility (NTF). The NTF, planned in the 1970s, represented one of the largest jumps in Reynolds number capability since the 1950s.[6] Many of NASA's older tunnels continued to put in good service, but they did not always provide accurate simulations, especially in the notoriously complex transonic region. Both the U.S. Air Force and NASA pursued plans for high–Reynolds number transonic tunnels in 1973–74, but as costs grew, their plans merged into the NTF. Technically, the NTF represented the state of the art in cryogenic transonic tunnels. It was designed to operate either with ambient air or pumped full of nitrogen in order to achieve full-scale Reynolds numbers.[7]

---

4. Ibid., p. 404.
5. Hallion, *On the Frontier*, pp. 252–258; Rumerman, *NASA Historical Data Book*, vol. 6, p. 383.
6. The Reynolds number is the ratio between inertial forces and viscous forces. With respect to wind tunnels, test data will more closely approximate actual conditions if the Reynolds numbers are the same for both. For a scale model to return similar data to that of a full-scale model, conditions in the wind tunnel must be altered. This is done sometimes by changing the velocity of the air or, as in the NTF, changing the atmosphere of the tunnel.
7. Donald D. Baals and William R. Corliss, *Wind Tunnels of NASA* (Washington, DC: NASA SP-440, 1981), pp. 133–134; NASA, *Aeronautics and Space Report of the President, 1988 Activities* (Washington, DC: NASA, 1988); Comptroller General of the United States, Report to the Committee on Science and Technology, House of Representatives, "Acquisition and Utilization of Wind Tunnels by the National Aeronautics and Space Administration (PSAD-76-133, GAO #090807, 23 June 1976); Comptroller General of the United States, Report to the Committee on Appropriations, House of Representatives, "The National Aeronautical Facilities Program: Issues Related to Its Cost and Need" (LCD-75-329, GAO #099407, 23 March 1976).

The 1980s witnessed a shift in tunnel operation, with increased use of contractors for both daily operations and renovations. For example, outside contractors performed the expansion of the 40-by-80-foot Ames tunnel to accommodate the 80-by-120-foot open circuit. Since the days of the NACA, in-house technicians had operated, calibrated, designed, and modified the Agency's wind tunnels. From the researchers' perspective, in-house staff gave the tunnels a kind of laboratory memory, people who knew the ins and outs of a tunnel, as well as test methods. Langley's slotted transonic tunnel innovation would hardly be conceivable without such an intimate understanding of the tunnel's characteristics. Certainly, outsourcing sought to reduce costs by reducing the number of technical personnel (such as shop workers who fabricated models). From 1979 to 1988, the number of NASA technical workers (which would have included wind tunnel technicians) decreased Agency-wide from 3,306 to 2,372. Conversely, the number of contractors increased in the same time period from 19,952 to 29,401. Within aeronautics, Ames and Lewis retained most of their technical support personnel, while Langley dropped from 1,056 to 909.[8] Tied to administrative changes that called for close accounting of tunnel time (i.e., no more free tunnel time for unfunded projects), the administrative layers between researchers and access to their experimental apparatus grew.

The market context for aeronautics research in the 1980s saw two shifts: one was the consequence of the Carter administration's decision to deregulate the airlines, and the second was the realization that European airframe manufacturer Airbus was a potent competitor. Deregulation allowed airlines to drastically alter their route structures. The old trunk lines swapped many of their point-to-point flights in favor of hub-and-spoke route networks. Airlines dropped service to unprofitable cities while increasing the frequency of service to competitive markets. The result was an increase in traffic at the hubs as well as an increase in the number of smaller jets. Where the early 1970s had seen the development of medium-range, wide-body aircraft such as the DC-10 and L-1011, deregulation favored smaller aircraft such as the Airbus A320 and the Boeing 737, as well as a raft of regional feeder aircraft (both jet and turboprop). Increased competition lowered ticket prices and led to numerous bankruptcies. The storied lines of Pan Am and Eastern Airlines both suffered in the 1980s and closed early in the next decade. Unable to rely on steady income from dedicated, regulated routes, American airlines moved into an era of greater financial insecurity. They sought new ways to cut costs, from reducing crew levels (Boeing eliminated the flight engineer from its Boeing

8. Rumerman, *NASA Historical Data Book*, vol. 6, table 7-1, p. 468; table 7-15, p. 480.

767 wide body) to building fleets that simplified maintenance and training (e.g., by purchasing only one type or family of aircraft). Airlines attempted to shift their fleets to more economical twin-engine aircraft, and the FAA assisted through Extended Twin-engine Operating Performance Standards (ETOPS), which allowed proven equipment and operators to fly greater distances from emergency landing sites. The long-term impact of deregulation for NASA was largely the challenge posed by a more crowded airspace, a concern that, as we have seen, began in the late 1960s.[9]

Ironically, deregulation conveniently meshed with Airbus's growth plans. The efficient A320 reached the market in time to feed a growing demand for twin-engine narrow bodies. While Airbus's earlier aircraft, the A300 and A310, were finally becoming viable product lines, the A320 was a smash hit. Airbus's growing market share came at the expense of American companies, notably McDonnell Douglas and Lockheed (Lockheed built its last commercial jet in 1983). The competitive threat from Airbus served as a rhetorical tool in Washington, DC, where it was mentioned not only as a reason for continued aeronautical R&D, but also as an example of the fruits of European industrial investment. The argument went that if Washington refused to respond to the European Union's (EU's) support for aviation, then the logical outcome would be the continued deterioration of a treasured American industry.[10]

The Office of Science and Technology Policy (OSTP), in its 1982 "Aeronautical Research and Technology Report," highlighted the threats posed by both Airbus and the Soviet Union.[11] In 1987, the OSTP reiterated its concern

---

9. Roger E. Bilstein, *Flight in America* (Baltimore: Johns Hopkins University Press, 2001), pp. 291–292, 343–346; Crouch, *Wings*, pp. 601–606.
10. Within the President's annual Aeronautics and Space Report, concern about Airbus appears in 1987: "Since the late 1940's, the United States has been the world leader in the development and application of aviation technology.... Today, however, U.S. technology is being challenged by friendly foreign competitors—the European Fighter Aircraft; Rafale, and A320 Airbus—as well as adversaries—the Soviet Su-27 and MiG-29 aircraft." NASA, *Aeronautics and Space Report of the President, 1987 Activities* (Washington, DC: NASA, 1987), p. 52. The following year, the report noted, "In recent years, however, the U.S. dominance in commercial transportation has been challenged by foreign competitors. Foreign aircraft now dominate the feeder airline/ medium-size aircraft market and significant thrusts have been made into major commercial airline markets, especially by the European consortium's Airbus series of transport aircraft." NASA, *Aeronautics and Space Report of the President, 1988 Activities* (Washington, DC: NASA, 1988), p. 45.
11. Executive Office of the President, Office of Science and Technology Policy, "Aeronautical Research and Technology Report," vol. 1, "Summary Report" (November 1982), p. 16.

over foreign competition, noting that "the Committee believes that the depth of foreign aeronautical resolve and the concerted national effort required to preserve American competitiveness are still largely underestimated."[12] Despite the emphasis on both military and commercial threats, the major thrust of aeronautical development in the decade was new military capabilities. The National Aero-Space Plane appeared to be a large-scale commercial project, but this was a fairly thin veneer. In fact, executive-level concern for commercial aviation appears to have had little impact on NASA's research programs. By way of contrast, the concern about noise, pollution, and energy efficiency in the Nixon and Carter presidencies was met with vigorous research packages. In the end, the actual research conducted by NASA in the 1980s reflected continued funding of traditional strands of research, as well as consistent funding of experimental military technologies. What is particularly intriguing about the top-level policy decisions is that they did little to change the *direction* of technical development. Though NASA's aeronautics research continued to be wide-ranging, many of the major leaps came from enabling technologies that had been conceived in the 1960s and developed in the 1970s. The impact of digital electronics was the most thorough, making possible radical aircraft configurations, redefining the design and testing process, and providing new tools for aircraft and airspace control. Similarly, NASA would harvest the fruits of its work in composites, laminar flow, and supercritical airfoils.

Numerical Aerodynamic Simulation

Until Ames acquired the Illiac IV, Harvard Lomax's CFD branch had been using commercially available computers, mostly from IBM, and writing their own top-level code. The Illiac IV was a departure because it took Ames into the business of piecing together a complex supercomputer. Not only did the researchers have to write code, they had to bring the machine through the development process; in essence, the Ames group was drawn into the problems of both software and hardware design. Though the Illiac IV's equipment was in place in 1972, it was a few more years before the CFD branch had the software to make the machine operate. As noted in the previous chapter, the Illiac IV did accomplish new types of aerodynamic modeling, but more than anything else, it served as a training ground for the Ames researchers.

Even as the Illiac IV was coming on line, it was clear that advances in integrated circuits and memory were such that CFD would soon benefit

12. Executive Office of the President, Office of Science and Technology Policy, "National Aeronautical R&D Goals: Agenda for Achievement," 1987, White House, Office of Science and Technology Policy, folder 12466, NASA HRC.

from having more advanced equipment. So in spite of having established branches in CFD and applied CFD at Ames, one of the keys to remaining on the forefront was obviously staying on the leading edge of computing equipment. This had been the case with wind tunnels; the NACA and NASA had continually sought to refresh their test equipment. The challenge was more acute for computing equipment, however, due to the field's faster pace of technological change. A wind tunnel could have a life of a few decades, especially if it received periodic upgrades, but a computer system was old in less than a decade. In 1975, even before the Illiac IV was fully operational, Victor Peterson, Bill Ballhaus, and Dean Chapman "dreamed up" a following act, the Navier-Stokes Processing Facility.[13]

By the end of the 1970s, CFD could model inviscid flows around simple objects. (An inviscid flow is one in which fluid has no viscosity.) Companies were beginning to use CFD codes to model their aircraft wings, Boeing on its 757 and Airbus on its A310. Airbus's efforts in CFD were noted as a competitive threat to American industrial CFD efforts. What remained was the much more demanding modeling of flow around complex objects and the modeling of viscous/turbulent flows. The Navier-Stokes Processing Facility Project (named after the fundamental equations that describe fluid dynamics) was to have a computing capability "three to four orders of magnitude" greater than the Illiac IV. One obvious lesson from the Illiac IV was that future computers needed to strike a balance between being advanced (esoteric devices that were developmentally challenging) and useful to researchers carrying out simulations. The Navier-Stokes computer was to be "user oriented and easy to program." Additionally, program managers had to be careful about building a machine that was too experimental; "development risk had to be low." The hardware portion of the facility was to be contracted out, while Ames's researchers would address the computer's compiler and programming. NASA endorsed the project, in principle, and its name became the Computational Aerodynamic Design Facility.[14]

Ames issued requests for proposals (RFPs) in 1976 and awarded initial design contracts to Burroughs and the Control Data Corporation. The name changed again to the Numerical Aerodynamic Simulation (NAS) facility in 1977. The Ames group also began building a larger coalition to support the project, visiting the large airframe manufacturers to brief them on the project

13. Victor L. Peterson, interview by Robert G. Ferguson, tape recording, Los Altos, CA, 17 January 2005, NASA HRC.
14. NASA Office of Aeronautics and Space Technology, "The Numerical Aerodynamic Simulator—Description, Status, and Funding Options," December 1981, folder 8740, NASA HRC.

and get feedback. NASA Headquarters also became interested in expanding the project's scope so that it could also serve non-aeronautical purposes. The Agency considered, for example, the possibility of using the NAS for climate modeling. By 1978, the Ames team began to consider a more capable computer, one that could both model the "performance of relatively complete aircraft configurations" and examine the physics of turbulence. As specifications and potential users multiplied, Ames established a User Steering Group that included representatives from industry, academia, and other government agencies. Ostensibly there to provide feedback, the Users Group also added to the NAS's inertia.[15]

The NAS's planning in the late 1970s and early 1980s occurred just as supercomputers emerged as a topical item. Seymour Cray, the computer designer who had been behind Control Data Corporation's (CDC's) fastest computers (and before that at Engineering Research Associates), had formed his own company and was producing the Cray-1. It was, depending on the metric used, about as fast as the Illiac IV and much more reliable. Ames bought a Cray-1, as did other government agencies and a number of aircraft manufacturers. Cray was becoming a household word. Japan added heat to these developments by subsidizing multiple supercomputer programs in an effort to take the world lead. After making substantial inroads into the American automobile market, Japan's threat to do the same for supercomputing was taken seriously in Washington, DC. NASA did not lose the opportunity to highlight the role of the NAS in keeping the United States competitive. Soviet advances in supercomputing added a Cold War element, although it was generally believed that the Soviets remained years behind.[16]

In the midst of this, it was becoming clear that the technology for the NAS computer did not yet exist, though it was "theoretically sound." The computer itself was becoming an R&D project. Ames established a Project Office in 1979, and Headquarters reclassified the computer as an R&D project rather than a simple facilities project. This was a computer so complex that researchers considered having a system that would simulate the computer. As the list of potential users grew, there was the question of how a machine narrowly designed around parallel processing might serve more general computing needs. The parallel processing architecture required researchers to compose their problem as discrete mathematical threads that could be handled simultaneously by multiple processors. This worked for aerodynamics and some other

15. Ibid.
16. See, for example, Peter J. Schuyten, "Soviet Gaining in Computers," *New York Times* (6 November 1980): D2.

areas that used iterative functions (such as computational chemistry), but it raised questions about how many scientists would and could actually use it.[17]

Finally, the initial Burroughs and CDC plans fell through when both companies reported that they could not meet the system requirements for the proposed price of $100 million. At the same time, commercially available supercomputers such as the Cray were within an order of magnitude of the initial NAS specifications. Rather than scrap the project, the NAS group reformulated their plans: "It became clear that a more reasonable approach was to build a system that was consistent with computer company commercial lines and flexible enough to use the best high speed computer available." The group explained that "when NAS was proposed it was intended to leap from the industry, but they did not anticipate how rapidly the industry would move. NASA didn't want to incur a high cost for an obsolete system."[18] The NAS became machine-independent, defined only by the promise of having the most advanced supercomputing capability commercially available.[19]

By 1983, the Ames group had proposed purchasing a single-processor Cray-2 that was to be operational by 1986. The plans were approved early in the year, and the NAS received funding in 1984. The Cray-2 was finally operational in March of 1987, billed as the most powerful supercomputer in the world at the time at 250 Mflops.[20] It was accessible from 27 remote locations and cost under $100 million. Designed to be scaled up over time, the system made use of Silicon Graphics, Inc. (SGI), workstations. SGI, a nearby computer company, was the most visible Silicon Valley beneficiary of Ames's computer research; the two organizations would work closely for the next decade, especially in the area of computer visualization.[21] By 1988, the NAS was installing an even faster Cray Y-MP. Operational in 1989, it offered sustained speeds of one billion calculations per second, thus finally meeting the original specifications set nearly a decade and a half earlier. By this time, however, the researchers'

17. Gina Bari Kolata, "Who Will Build the Next Supercomputer?" *Science* 211 (16 January 1981): 268–269.
18. "NAS Briefing to Tony Taylor and Harriett Smith, 8/27/82," memo for record, Computers 1970–1982, folder 8740, NASA HRC.
19. Benjamin M. Elson, "New Computers Will Aid Advanced Designs," *Aviation Week & Space Technology* (29 August 1983): 50–57.
20. *Mflop* stands for "one million floating point operations per second."
21. NASA News, "Fact Sheet: NASA Numerical Aerodynamic Simulation Facility," press release no. 87-05, 12 February 1987, folder 8738, NASA HRC.

appetite for computational power had grown, and they were planning for one trillion calculations per second before the year 2000.[22]

The acquisition of the Illiac IV in 1972 appeared to be a coup for Ames, but it was ultimately a mixed blessing. As early as 1977, Sy Syvertson wrote that "our years of difficult learning experience with the Illiac IV parallel processor, and our experience in developing efficient software for parallel processing will be invaluable in carrying NAS Facility into successful operation."[23] One of the legacies of the Illiac IV was an inclination toward inherently risky experimental computing hardware rather than commercially available equipment. Ames wanted to be one or two steps ahead of everyone else, but that meant developing new computer hardware faster than the marketplace. It was not until 1983, after initial design studies and a significant amount of campaigning, that the NAS group abandoned the custom-built approach. In the meantime, other laboratories and companies were populating their computing facilities with Cray supercomputers. Indeed, by 1986, Lewis had its own Cray X-MP, a machine that was almost as fast as the forthcoming Cray-2 at Ames.[24] The custom-built approach appears, in hindsight, as an expensive and protracted diversion, despite its merits as a learning tool in the 1970s. Ultimately, what distinguished the Ames CFD efforts was not its equipment (which, increasingly, many labs began to replicate) but the core of theorists, mathematicians, and fluid dynamicists who turned out innovative CFD solutions regardless of the vicissitudes of the NAS.

Unconventional Wings: The AD-1, RSRA, and AFTI-MAW

From December 1979 to August 1982, a strange jet aircraft took flight from Dryden with a wing that pivoted about its center. At takeoff, the aircraft's wing was perpendicular to the fuselage, but at altitude, the pilot rotated the wing such that the right-hand side swept forward and the left-hand side swept backward. This was the AD-1, the world's first and, to date, only piloted oblique wing aircraft. In spite of the aircraft's unusual appearance, Robert T. Jones, the Ames aerodynamicist who had nurtured the concept of oblique wings, saw the configuration as a viable supersonic commercial aircraft. This did not come to pass, and although Jones retired from NASA in 1981, he continued to advocate

22. NASA News, "World's Fastest Supercomputer Installed at NASA," press release no. 88-152, 10 November 1988, folder 8738, NASA HRC.

23. C. A. Syvertson to Dr. Alan M. Lovelace, 12 October 1977, OAST Chron Files, "NASF," folder 18274, NASA HRC.

24. NASA News, "New Super Computer Supports Key NASA Projects," press release no. 86-74, 11 June 1986, folder 8738, NASA HRC.

oblique wings, including oblique flying wings. More recently, DARPA renewed interest in the technology with a $10 million Northrop Grumman design study of supersonic oblique wings.[25]

The oblique wing concept originally emerged in World War II in an unrealized German experimental design, the Blohm & Voss P.202. Researchers at Langley briefly experimented with the idea in 1946 using a small, electrically controlled free-flight model.[26] It was Jones who ultimately nurtured the concept, initially playing with the idea theoretically in the 1950s. Over time, he took the idea more seriously and, by the 1970s, believed that it was a practical possibility that had significant advantages over competing designs.[27] The closest existing configuration was the variable sweep wing, or swing wing, which angled both wings backward. Swing wings, like those on the General Dynamics F-111 or North American Rockwell B-1, by virtue of their ability to move from a straight-wing to a swept-wing configuration, sought good performance at both low and high speeds. Unfortunately, variable sweep came at great cost. The wings required heavy hinges and actuators. Moreover, swing wings changed the aircraft's center of gravity and aerodynamic center as the wings swept back; the required trimming added drag. The oblique wing offered a simpler structure, a simpler and lighter pivoting mechanism, and less need for fore and aft trimming.[28] The engineering challenges of oblique wings, however, were patently obvious. Special attention would need to be paid to the control of the aircraft as well as the aeroelastic performance of the forward-swept wing.

Jones's group at Ames began wind tunnel and flight testing in earnest in the 1970s. Initial results showed that oblique wings were a viable configuration for transonic and low-supersonic speeds.[29] In the wake of the SST cancella-

---

25. DARPA, "DARPA Begins Unique Oblique Flying Wing Program," press release, 17 March 2006.
26. Edward C. Polhamus and Thomas A. Toll, "Research Related to Variable Sweep Aircraft Development" (Langley Research Center, Hampton, VA: NASA TM-83121, May 1981); Michael J. Hirschberg, David M. Hart, and Thomas J. Beutner, "A Summary of a Half-Century of Oblique Wing Research" (45th AIAA Aerospace Sciences Meeting and Exhibit, Reno, NV, 8–11 January 2007).
27. Walter G. Vincenti, "Robert Thomas Jones, 1910–1999," *Biographical Memoirs*, vol. 86 (Washington, DC: National Academies Press, 2005).
28. Alex G. Sim and Robert E. Curry, "Flight Characteristics of the AD-1 Oblique Wing Research Aircraft," (Dryden Flight Research Facility, Edwards, CA: NASA TP-2223, March 1985).
29. Robert T. Jones, "Properties of Oblique-Wing/Body Combinations for Low Supersonic Speeds," *Vehicle Technology for Civil Aviation* (1 January 1971): 389–407; Lawrence A. Graham, Robert T. Jones, and Frederick W. Boltz, "An Experimental Investigation of an Oblique-Wing and Body Combination at Mach Numbers Between 0.60 and 1.40" (Ames Research Center, Moffett Field, CA: NASA TM-X-62207, December 1972).

tion, Jones saw the oblique wing as a candidate for reviving the development of high-speed commercial transports and championed it as such.[30] The canceled SST was a swing-wing design before weight problems pushed Boeing to revert to a fixed-wing configuration.[31] The oblique wing held out the hope of providing a variable sweep wing without the swing wing's weight problems. Additionally, model tests showed that oblique wings generated lower wave drag at transonic speeds. The configuration received additional support from a NASA-sponsored design analysis performed by Boeing from 1972 to 1973, in which the company examined various high-transonic configurations up to Mach 1.2. In this speed range, sonic booms refract in the atmosphere and do not reach the ground, thus circumventing one of the factors that doomed the original SST. Boeing compared five different designs: a fixed swept wing, a variable sweep wing, a delta wing, an oblique wing ("yawed" in Boeing parlance) with twin fuselages, and an oblique wing with a single fuselage. Boeing's engineers concluded that only the single-fuselage yawed-wing design could achieve Mach 1.2 and meet future noise restrictions.[32]

Throughout the 1970s, the Ames group pushed forward with a varied set of oblique-wing studies. They made wind tunnel studies of an F-8 equipped with an oblique wing; the F-8 was the same aircraft used for the supercritical airfoil studies.[33] Calculations of the configuration's noise showed that it produced less sonic boom overpressure than swing wings.[34] By mid-decade, Ames had fielded the Oblique Wing Remotely Piloted Research Aircraft. The aircraft flew three times in 1976 and exhibited the ability to pivot its wing to a 45-degree angle.[35] Meanwhile, under contract to NASA, Boeing continued to refine its

---

30. Robert T. Jones, "New Design Goals and a New Shape for the SST," *Astronautics and Aeronautics* (December 1972): 66–70.
31. Horwitch, *Clipped Wings*, p. 188.
32. Robert M. Kulfan et al., "High Transonic Speed Transport Aircraft Study, Final Report" (prepared by the Boeing Commercial Airplane Company for the Ames Research Center, Moffett Field, CA: NASA CR-114658, September 1973).
33. Lawrence A. Graham, Robert T. Jones, and James L. Summers, "Wind Tunnel Tests of an F-8 Airplane Model Equipped with an Oblique Wing" (Ames Research Center, Moffett Field, CA: NASA TM-X-62273, June 1973).
34. R. M. Hicks and J. P. Mendoza, "Oblique-Wing Sonic Boom" (Ames Research Center, Moffett Field, CA: NASA TM-X-62247, February 1973).
35. "Oblique Wing Remotely Piloted Research Aircraft, vol. 1: Development" (NASA CR-114723); R. Dale Reed, "Flight Research Techniques Utilizing Remotely Piloted Research Vehicles" (Advisory Group for Aerospace Research and Development [NATO] Aircraft Assessment and Acceptance Testing, 1 May 1980).

(NASA image ECN-15846)

Figure 5.1. AD-1 exhibits its full 60-degree wing sweep in flight.

oblique-wing transonic transport design, narrowing design and performance parameters. The oblique-wing arrangement presented some challenges in terms of where to locate the engines and the landing gear since these could not be attached to the wing without pivoting them as well. In 1977, Boeing presented NASA with a final configuration.[36]

The Boeing report provided important design inputs to what would become the AD-1, a small, low-speed, low-cost, piloted experimental vehicle. The AD stood for Ames-Dryden. The aircraft had a wingspan of just over 32 feet and could pivot its wings 60 degrees from perpendicular. Similar to Boeing's recommended design, the AD-1's engines and landing gear were mounted on short fuselage extensions just aft of the wing pivot. The Ames Industrial Corporation of Bohemia, New York, manufactured the aircraft for $240,000. From 1979 to 1982, the aircraft made 79 flights from Dryden. Although the aircraft was not designed to fly faster than 170 miles per hour, it was able to vary its wing sweep in flight and provide valuable test data on aerodynamic and control issues.[37]

36. Boeing Commercial Airplane Company Design Department, "Oblique Wing Transonic Transport Configuration Development" (NASA CR-151928, January 1977).

37. Alex G. Sim and Robert E. Curry, "Flight Characteristics of the AD-1 Oblique-Wing Research Aircraft" (Ames Research Center, Dryden Flight Research Facility, Edwards, CA: NASA TP-2223,

The results of the AD-1 investigation generated interest among researchers, but there were no following acts. There was an Oblique Wing Research Program in the mid-1980s that planned to fit an oblique wing to one of Dryden's F-8 aircraft for a first flight in 1991, but this never panned out.[38] The idea did not wholly disappear, however. In the late 1980s, R. T. Jones was instrumental in reviving an earlier version of his oblique-wing ideas, namely, an oblique flying wing. There would be no fuselage in such an aircraft. Crew and passengers were to sit inside the wing, and the engines would pivot with the wing's sweep. The public unveiling of the Northrop B-2 flying wing in 1988 made an oblique version that much more conceivable.[39] Ames continued to support basic work on the idea through the 1990s, performing conceptual design studies in conjunction with Stanford University, subcontracting studies to Boeing and McDonnell Douglas, and performing wind tunnel analysis in the Ames 9-by-7-Foot Supersonic Wind Tunnel. These tests showed that the hoped-for cruise performance was not as good as expected and that there was insufficient basis for estimating the weight of the aircraft, a necessary step in determining the economic viability of the idea.[40] A decade later, DARPA resumed the effort, awarding Northrop Grumman a contract to study the technology and begin preliminary design work on a supersonic, tailless, oblique flying wing.[41]

NASA fielded another unusual research aircraft in the 1980s, the Rotor Systems Research Aircraft (RSRA). The RSRA had strong similarities to the

---

March 1985); Robert E. Curry and Alex G. Sim, "In-Flight Total Forces, Moments, and Static Aeroelastic Characteristics of an Oblique-Wing Research Airplane" (Ames Research Center, Dryden Flight Research Facility, Edwards, CA: NASA TP-2224, October 1984).

38. Glenn B. Gilyard, "The Oblique-Wing Research Aircraft: A Test Bed for Unsteady Aerodynamic and Aeroelastic Research," in *Transonic Unsteady Aerodynamics and Aeroelasticity*, part 2 (symposium held in Hampton, VA, 20–22 May 1987; Langley Research Center, Hampton, VA: NASA Conference Publication 3022, 1989).
39. Alexander J. M. Van der Velden, "Conceptual Final Paper on the Preliminary Design of an Oblique Flying Wing SST" (NASA CR-182879, December 1987); Alexander J. M. Van der Velden, "The Conceptual Design of a Mach 2 Oblique Flying Wing Supersonic Transport" (Ames Research Center, Moffett Field, CA: NASA CR-177529, May 1989).
40. Thomas L. Galloway et al., "Large Capacity Oblique All-Wing Transport Aircraft," in *Transport Beyond 2000: Engineering Design for the Future* (workshop held 26–28 September 1995; Langley Research Center, Hampton, VA: NASA Conference Publication 10184), pp. 461–489.
41. DARPA, "DARPA Begins Unique Oblique Flying Wing Program," news release, 17 March 2006; Michael J. Hirschberg, David M. Hart, and Thomas J. Beutner, "A Summary of a Half-Century of Oblique Wing Research" (45th AIAA Aerospace Sciences Meeting and Exhibit, Reno, NV, 8–11 January 2007).

XV-15 tilt-rotor in terms of research focus and organizational foundations. The vehicle was intended to serve as a kind of flying laboratory for different technical approaches, but it was, like the XV-15, attempting to bridge the gap between helicopters and winged aircraft. Organizationally, the RSRA was a joint Army-NASA project, initially operated out of Langley beginning in 1970. Sikorsky built two RSRAs and delivered them to Langley in 1979. The two vehicles were soon transferred to Ames as the Center became the lead for rotorcraft and power-lift research.[42]

The initial impetus for the RSRA was to have a flying rotorcraft laboratory. This goal grew out of the difficulty of accurately predicting the performance of new rotor technologies without actually flight-testing them. It had been the case in rotor design that manufacturers usually had to modify helicopter airframes for each rotor variant. The RSRA was designed to accept different rotors and serve as a "flying wind tunnel," providing capabilities that even a wind tunnel could not (e.g., high-g maneuvers). One of the features of the RSRA was that it could operate in three different modes: pure helicopter mode, compound mode with lift shared by rotors and a fixed wing, and solely as a fixed-wing aircraft. This last mode existed primarily for safety reasons should the pilots need to jettison their rotors in the event of an emergency.[43] Mechanically, the most important feature of the RSRA was an "active-isolator" rotor balance system that isolated rotor vibrations and measured rotor forces and moments. This was the element that made the RSRA truly a flying laboratory. As with most sensitive laboratory measurement devices, the RSRA had to be carefully calibrated.[44] Unfortunately for the RSRA, the vehicle took many years to fine-tune (for both technical and budgetary reasons), and even as late as 1986, only one of the vehicles had reached "near-operational" status.[45]

The second RSRA became subsumed by what was called the X-wing experiment. In the early 1980s, DARPA, NASA, and the Navy became interested in the possibility of a vehicle that would take off like a helicopter and then

42. David D. Few, "A Perspective on 15 Years of Proof-of-Concept Aircraft Development and Flight Research at Ames-Moffett by the Rotorcraft and Powered-Lift Flight Projects Division, 1970–1985" (Ames Research Center, Moffett Field, CA: NASA Reference Publication 1187, 1987); Rumerman, *NASA Historical Data Book*, vol. 6, pp. 205–208.

43. W. D. Painter and R. E. Erickson, "Rotor Systems Research Aircraft Airplane Configuration Flight-Test Results" Ames Research Center, Dryden Flight Research Facility, Edwards, CA: NASA TM-85911, 1984).

44. C. W. Acee, Jr., "Final Report on the Static Calibration of the RSRA Active-Isolator Rotor Balance System" (Ames Research Center, Moffett Field, CA: NASA TM-88211, 1987).

45. Few, "A Perspective on 15 Years."

transition to horizontal flight by stopping an X-shaped rotor. The rotor blades had a symmetrical airfoil and generated both lift and directional control by blowing engine exhaust through leading and trailing edge slots in the blade. The idea worked when tested in the Ames 40-by-80-foot tunnel using a scale rotor. Sikorsky modified the RSRA to accept the X-wing rotor and delivered the vehicle in 1986 to Ames-Dryden. The X-wing made it as far as taxi tests before the program was canceled entirely in 1988 due to rising costs and complexity.[46]

The Advanced Fighter Technology Integration (AFTI) program was a joint program that included Dryden and the U.S. Air Force. The program's first aircraft was an F-111 swing-wing aircraft, which was used in the 1970s for transonic supercritical wing studies but modified with a wing that could change shape. Rather than a simple arrangement of leading-edge slats and trailing-edge flaps, the Mission Adaptive Wing (MAW), as it was called, created a smooth but alterable camber using rotary actuators and flexible fiberglass skins. Boeing designed and built the wing beginning in 1979, and the MAW first flew in October 1985. The first phase of testing employed manual control of the wing; the second phase saw the addition of computers that would automatically adjust camber. In total, the aircraft flew for 145 hours of testing, with the final flight in 1988. The MAW increased the F-111's range and maneuverability. The MAW's proponents at Dryden and Boeing argued that the variable camber system was no heavier than a traditional flap and slat system. Further, they argued that because the entire works were enclosed within the fiberglass panels, the wing would require less maintenance than traditional exposed flap and slat systems.[47]

Advanced Turboprop Project

The Advanced Turboprop Project (ATP) became a part of the Aircraft Energy Efficiency Program (covered in the previous chapter), even though it predated the ACEE by a number of years. In fact, the ATP was one of Lewis's first post-Apollo bids to return to air-breathing engines. In this respect, ATP was a key technology for reinvigorating the Ohio laboratory, but the ATP had a long gestation period. Even after it moved into the testing phase in the 1980s, it failed to move beyond the experimental stage. Historians Mark Bowles and Virginia Dawson have argued that fuel price variations were critical to the technology's

46. Ibid.; Rumerman, *NASA Historical Data Book*, vol. 6, pp. 205–208.
47. John W. Smith, Wilton P. Lock, and Gordon A. Payne, "Variable-Camber Systems Integration and Operational Performance of the AFTI/F-111 Mission Adaptive Wing" (Edwards, CA: NASA TM-4370, 1992); Scott E. Parks (Major, U.S. Air Force), "Pilot Report: AFTI F-111" (Air Command and Staff College, Report Number 88-2050).

adoption, that ATP was born in an era of high fuel prices and then stagnated when prices dropped.[48] Certainly fuel prices were a significant factor, but the ATP represented a risky new technology that could not be implemented in an incremental fashion. In the case of composite aircraft materials, for example, airlines slowly introduced the materials over a period of three decades, gradually increasing the percentage of composites to the point that Boeing's 787 is mostly composite. To adopt the ATP would be to assume a large and discrete risk. Additionally, the ATP was competing against the highly refined (and known) technical context of jet engines. So while the ATP did show worthwhile efficiency gains in experimental testing, the commercial risks were daunting.

The ATP project sought to create a turboprop that could compete against subsonic jet engines. The idea was to take the core of a jet engine and, rather than have a high-bypass fan section (the blades that one sees when looking at the front of a modern jet engine), instead turn a large propeller with wide, swept blades. It was these swept blades that provided the initial inspiration for the project. In the early 1970s, Lewis engineer Daniel Mikkelson found that one solution to the problem of propeller blade compressibility was to sweep the blades.[49] Compressibility was one of the classic problems of propellers and a principal reason for the search for an alternative means of propulsion in the 1930s and '40s. As propeller blades spin faster and faster, the tips reach transonic speeds, encountering steep increases in drag as well as harmful turbulence. Sweeping the blades accomplishes the same thing that sweep does for wings: it delays the onset of these effects. This encouraged Lewis engineers to consider whether such a propeller could be used not just as a replacement for existing turboprop propellers, but as part of an engine that could reach speeds comparable to those attained by existing jet engines.

While the ATP project formed part of the original set of ACEE proposals made by NASA to Congress, the turboprop program did not receive funding initially.[50] In lieu of support from Congress and Headquarters, the Center gave provisional funding to measure the efficiency of a scale model in a wind tunnel. With these results as proof of the concept's promise, the Lewis group won funding in 1978. For the Lewis researchers, the ATP was their first experience

48. Mark D. Bowles and Virginia P. Dawson, "The Advanced Turboprop Project: Radical Innovation in a Conservative Environment," chap. 14 in *From Engineering Science to Big Science: The NACA and NASA Collier Trophy Research Project Winners*, ed. Pamela E. Mack (Washington, DC: NASA SP-4219, 1998).

49. Ibid.

50. NASA News, "Senate Group Briefed on Aircraft Fuel-Saving Effort," release no. 75-252, 10 September 1975, folder 18273, NASA HRC.

managing a large program that drew together the other Centers and industrial subcontractors, which included Pratt & Whitney, Allison, and Hamilton Standard. The initial phase of the project examined four problems: propeller design, noise, installation (how to attach such an engine to an airframe), and drive systems (how to attach the propeller to the modified jet engine core). By 1981, Hamilton Standard (a longtime propeller manufacturer) won the contract to design a composite propeller (renamed a propfan). The following year, General Electric began its own research on an alternative propfan (eventually called the GE-36, an unducted fan), one that eliminated the need for a gearbox by running the propeller blades straight off a low-speed turbine stage. Headquarters directed the Lewis group to cooperate with GE.[51]

In 1986, GE had completed its unducted fan prototype and begun flight tests. At the same time, Hamilton Standard was finally delivering its 9-foot-diameter prototype. Lacking a tunnel large and fast enough for testing in the United States, NASA took the propfan to Europe, to the ONERA facility in Modane, France. Next, researchers mated the propeller to an Allison engine and gearbox and performed ground tests in California. Finally, in 1987, they installed the system on one wing of a Gulfstream II (a twin-engine business aircraft) and proved that the system delivered the 20 to 30 percent fuel savings that Lewis engineers had promised more than a decade earlier. Unfortunately for Pratt & Whitney, both Boeing and McDonnell Douglas were more interested in the GE engine and were making preliminary plans to manufacture aircraft using the GE-36. These were to be the Boeing 7J7 and McDonnell's MD-91X/92X/94X and were to enter service in the early 1990s. McDonnell Douglas went so far as to test the unducted fan on an MD-80. Very quickly, however, the manufacturers canceled all such plans.[52]

For their efforts, the team from Lewis and their industrial partners won the Collier Trophy in 1987, an award given for "significant achievement in the advancement of aviation." Although it was the simpler General Electric design that the airframe manufacturers planned to install, the GE-36 was most certainly inspired by the work that began at Lewis in the early 1970s. Propfans came within a whisker of actually being adopted; they were shelved not just by falling fuel prices, but also a market context that pitted the new propfan designs against the forthcoming Airbus A320. The A320 was an advanced, fly-by-wire narrow-body design with efficient conventional jet engines. It gave the airlines good efficiency without entailing the unknown risks of propfans (e.g., actual life-cycle costs, noise). Boeing and McDonnell Douglas responded to

51. Bowles and Dawson, "The Advanced Turboprop Project."

52. Ibid.; NASA, *Spinoff 1987* (Washington, DC: NASA).

the A320 with similarly efficient versions of traditional configurations. Perhaps the strongest beneficiary of the ATP project was Hamilton Standard (now Hamilton Sundstrand), which went on to design and market swept composite propellers for conventional turboprop engines. One of the questions that the parallel GE-36 development raises is whether the ATP project was well suited to building a working prototype, especially given the inevitable bureaucratic delays in funding that often attend low-priority government projects. Conversely, how important was NASA's role in stimulating GE's research in the first place? Pratt & Whitney, for its part, took the technology in a different direction. In the 1990s, it pursued, with assistance from NASA, the Advanced Ducted Propulsor, the Advanced Technology Fan Integrator, and finally the geared turbofan (GTF), this last one being the most descriptive.[53] The GTF connected the low-pressure turbine shaft to a set of sun and planet gears that were then connected to enclosed fan blades. This reduced the rotational speed of the fan blades, and, although the gearbox added weight and complexity, it allowed engineers to optimize both turbine and fan speeds (similar to the ATP). Designed to provide ATP-like efficiency gains, it did so in an incremental fashion and without the noise and vibration concerns inherent in the ATP. As of the writing of this book, Pratt & Whitney was offering the GTF as the PW1000G for the narrow-body commercial transport market.[54]

Maneuverability and the Flight-Test Revival

In the 1980s, NASA and the Department of Defense reinvigorated experimental flight testing at a pace reminiscent of the heyday of the X-vehicles in the 1950s. This renaissance was partly the result of increased defense funding under President Reagan, but it also had its roots in a larger shift within airpower strategy and tactics. To review, the 1950s experimental program emphasized speed above all else and culminated in the X-15, a Mach 6 rocket plane. Within the military, the focus on speed reached a crescendo with the SR-71 (a Mach 3 reconnaissance aircraft), the XB-70 (a prototype Mach 3 bomber), and the F-104 (a minimalist fighter design that emphasized supersonic performance). What the services, and especially the U.S. Air Force, realized, however, was that speed alone could

53. NASA supported research at Pratt & Whitney in the 1970s that examined, among other things, geared turbofans. See G. A. Champagne (Pratt & Whitney Aircraft), "Study of Unconventional Propulsion System Concepts for Use in a Long Range Transport" (Lewis Research Center, Cleveland, OH: NASA CR-121242/PWA-4692, August 1973).
54. Daniel Solon, "Bumpy Ride Ahead for Air Industries," *New York Times* (18 July 2010): Business section. It should be noted that small-sized GTFs have been produced and operated since the 1970s by Garrett AiResearch and Lycoming (both now Honeywell) for use on business jets and small regional aircraft.

not overcome growing battlefield threats from improved radar and missiles and that there remained an important role for aircraft that could engage the enemy close in, i.e., in a dogfight. In the realm of strategic bombing, the province of the U.S. Air Force, high altitude and speed were no longer sufficient to keep the bombers out of the range of ground-to-air missiles. Instead, it became desirable to hide the aircraft from detection, such as through terrain-hugging flightpaths and stealth technologies. The North American Rockwell B-1 bomber, which President Carter canceled and President Reagan revived, was a transitional aircraft in this respect: although it was originally designed for Mach 2 flights, changes in enemy capabilities and doctrine (not to mention cost) resulted in an aircraft designed primarily for subsonic attack at nap-of-the-Earth altitudes.[55] The Lockheed F-117 stealth fighter and the Northrop B-2 stealth bomber both ignored high speed in favor of low observable technologies. For fighter aircraft, the experience of Vietnam illustrated the importance of aircraft maneuverability. The first example of this new emphasis was the General Dynamics F-16, developed in the 1970s as part of the Lightweight Fighter Program. By no means a slow aircraft, the F-16 gave pilots the ability to make high-G turns and thus maintain a tactical edge over enemy aircraft.[56]

From the late 1970s onward, NASA and the Department of Defense began a series of experimental flight tests that examined different aspects of high-maneuverability flight.[57] NASA's contribution was, as in earlier decades, the considerable scientific and engineering expertise necessary for framing test problems and analyzing results. Conversely, DOD established the primary research objectives and provided necessary funding. Technically, the maneuverability programs picked up and built on a number of strands of research that NASA had investigated in the 1970s. Among the most important was the use of digital flight control; most of these vehicles could not fly without computer mediation. The vehicles often used multiple control surfaces, which had to be adjusted, in concert, in fractions of a second. In this, Dryden had the most experience after its work with digital fly-by-wire, but the other Centers also were well represented. Ames and Langley supported the work with wind tunnel research. Langley assisted in

55. The redesigned B-1, though still a swing wing, was primarily a subsonic aircraft with a modest ability to fly supersonic.

56. Robert Corum, *Boyd: The Fighter Pilot Who Changed the Art of War* (New York: Back Bay Books, 2004); Verne Orr, "Developing Strategic Weaponry and the Political Process. The B1-B Bomber: From Drawing Board to Flight" (diss., Claremont Graduate University, 2005). Verne Orr was Secretary of the Air Force under President Reagan.

57. Stealth was the other major area of aeronautical R&D at this time. Since stealth was and remains a highly classified topic, the extent of NASA's involvement is not currently definitive.

aerodynamic and structural studies. Lewis brought to bear its understanding of jet engines and thrust vectoring, a technique that proved to be one of the most effective over the long run. Something that is, perhaps, not immediately clear from this second wave of experimental vehicles is that they represented a concerted effort to model unusual configurations and unusual maneuvers. Each vehicle was certainly an attempt to do something new, but behind the vehicles and their technical accomplishments, NASA researchers were developing modeling and testing techniques (i.e., wind tunnel and CFD) that would be generally applicable and useful for the design of future aircraft. An excellent example of this was the High Alpha Technology Program, begun in 1986, which correlated the modeling and flight testing of high-angle-of-attack research across Centers and vehicles. So while the test aircraft of this era were varied in their objectives, methodologies, and funding, behind them were consistent communities of researchers extracting long-term lessons about highly maneuverable vehicles.[58]

HiMAT/Remotely Piloted Research Vehicle

The Remotely Piloted Research Vehicle (RPRV) was a project that grew out of the clever use of remote control hobby aircraft at Dryden in the 1960s. Dale Reed, who championed lifting-body experiments, had bootstrapped his research through small-scale testing, including the use of remote control aircraft that he borrowed from his own hobbies. Using an oversized remote control hobby model, Reed carried scale lifting-body models to altitude and dropped them as gliders, in essence mimicking the piloted drop tests that had been standard practice at Dryden since the early rocket-plane experiments. By the late 1960s, Reed began exploring whether test pilots could fly experimental aircraft remotely using instruments, as opposed to the hobbyist practice of flying the model while looking skyward from the ground. Dale put the technology to work in his Hyper III lifting body (a Langley-derived design) that, on a single flight in 1969, was dropped from a helicopter and taken through its test flight by a pilot on the ground. For Reed and his colleagues, remote control and remote piloting offered an inexpensive way to vet designs and build momentum for further support. As the lifting-body experiments drew to a close, remote piloting appeared to be, in itself, a promising technique.[59] Dale Reed's group then outfitted the Center's Piper PA-30 Twin

58. See, for example, the proceedings of NASA's Fourth High Alpha Conference, NASA CP-10143, NASA Dryden Flight Research Center, 12–14 July 1994.

59. Hallion, *On the Frontier*, pp. 210–216 and appendix E (table 2); Dwain A. Deets, V. Michael DeAngelis, and David P. Lux, "HiMAT Flight Program: Test Results and Program Assessment Overview" (Dryden Flight Research Facility, Edwards, CA: NASA TM-86725, 1986).

Comanche with the necessary instrumentation for remote piloting. Initially using a safety pilot riding in the aircraft, the group eventually flew the plane entirely by remote from takeoff to landing. Concurrently with the work on the Piper, Reed's group collaborated with the Air Force in writing a proposal to test the spin characteristics of the forthcoming F-15 fighter aircraft using a scale, remotely piloted vehicle. It was approved in 1972, and Dryden flew two 3/8 scale models air-dropped from B-52s.[60]

The Dryden group succeeded in getting approval for a large-scale RPRV project mid-decade, jointly supported by NASA and the Air Force Flight Dynamics Laboratory at Wright-Patterson Air Force Base. The Highly Maneuverable Aircraft Technology, or HiMAT, was a scale, jet-powered aircraft that explored high-g maneuvers. North American Rockwell (the successor to North American Aviation) won the contract in 1979 to build two HiMAT vehicles. Flight testing began in 1979 and continued into early 1983 for a total of 26 flights. The HiMAT was dropped by a B-52 and landed on the dry lakebed on skids. It had a canard configuration, and the rear wings incorporated winglets. HiMAT succeeded in flying extremely tight turns with very high g-forces, enabled by what was called "relaxed static stability," or to put it another way, the company had designed a vehicle that was highly unstable, yet entirely controllable. Strong yet lightweight fiberglass and graphite materials (making up some 30 percent of the vehicle), as well as a digital fly-by-wire control system, allowed the HiMATs to remain intact and under control through maneuvers that were simply impossible in other high-performance aircraft of the time.[61]

In comparison to the General Dynamics F-16 aircraft, the HiMAT could fly high-speed turns in about half the radius. The HiMAT design also validated the use of composites and the canard arrangement. Both would become part of another high-maneuverability program, the X-29. As for the RPRV capability, remote piloting has become an important research option, and not just at Dryden. The reduced cost and risks associated with RPRV flights have allowed NASA to squeeze more out of its budget. The HiMAT also contributed a new test procedure that could be used for both piloted and unpiloted vehicles: a flight-test maneuver autopilot. This was a Teledyne Ryan Aeronautical autopilot that essentially took the test pilot out of the

60. Ibid.

61. Ibid.; Rumerman, *NASA Historical Data Book*, vol. 6, pp. 214–217; "NASA Facts: HiMAT Highly Maneuverable Aircraft Technology" (Dryden Flight Research Center, Edwards, CA: FS-2002-06-025-DFRC, 2002).

(NASA image ECN-14284)

Figure 5.2. The HiMAT remotely piloted vehicle attached to its B-52 mother ship prior to airdrop.

loop. Researchers could program a particular flight sequence and repeat or modify the test as necessary.[62]

Advanced Fighter Technology Integration/F-16

The AFTI program's second aircraft was an F-16 modified with advanced sensors, avionics, and flight control capabilities. In these tests, researchers used the aircraft's digital flight controls to optimize control laws for specific missions (e.g., air-to-air combat, air-to-ground attack). Engineers decoupled the flight controls and recombined them digitally such that the aircraft handled better than the production version. This kind of flight control would become even more important in future generations of high-performance aircraft. The program also included the testing of voice commands, the first time this had been done in developmental testing in a fighter aircraft. They achieved an overall voice recognition rate of 78 percent.[63]

---

62. Eugene L. Duke, Frank P. Jones, and Ralph B. Roncoli, "Development and Flight Test of an Experimental Maneuver Autopilot for a Highly Maneuverable Aircraft" (Dryden Flight Research Facility, Edwards, CA: NASA TP-2618, 1 September 1986).
63. Steven D. Ishmael and Capt. Donald R. McMonagle, "AFTI/F-16 Test Results and Lessons" (presented at the Symposium of the Society of Experimental Test Pilots, Beverly Hills, CA, 27

X-29 Forward-Swept Wing

The X-29 Forward-Swept Wing aircraft flew from 1984 to 1992. Like the HiMAT, it originated in the 1970s, born of the same interest in high-maneuverability fighter aircraft. Like HiMAT, the X-29 would explore an unusual and highly unstable configuration. Advanced composite structural design and digital flight control would likewise be critical enabling technologies. The X-29 was a joint DARPA–Air Force Flight Dynamics Laboratory project, but NASA was in charge of flight testing at Dryden. DARPA and the Air Force issued their RFP in 1977 and awarded Grumman the contract for two test vehicles in 1981.[64]

The X-29 was a jet aircraft with a canard configuration, the main wings sweeping forward at a 30-degree angle. Forward sweep was an old idea; Germany, in fact, had flown forward-swept prototypes in World War II. Forward sweep causes the airstream to flow in, toward the fuselage (the opposite of a swept-back wing). This meant that at extremely high angles of attack, when a traditional wing would lose controllability, the forward-swept wing retains airflow over the ailerons and thus remains controllable. The tradeoff is that the arrangement makes the aircraft highly unstable and the wings experience higher-than-usual bending and twisting loads. The X-29 used a digital fly-by-wire system to make the aircraft flyable. A computer managed the aircraft's unconventional control surfaces, which, in addition to a rudder, included full-motion canards (i.e., the entire canard rotated), flaperons (combined flap/ailerons located on the trailing edge of the main wing), and strake flaps (horizontal flaps located astride the rear fuselage). The digital fly-by-wire used all of the control surfaces in a coordinated fashion and did so far faster than a human ever could have. The X-29's forward-swept wing was specially designed to resist high aerodynamic forces. Engineers at Langley tested the X-29 design in the 16-Foot Transonic Tunnel in order to model the wings' aeroelasticity (i.e., how the wings bent, or would attempt to bend, under anticipated aerodynamic loads). With these data, Grumman crafted a wing from hundreds of carbon-Kevlar composite layers. The result was a lightweight wing that could bend and yet remain structurally and aerodynamically sound.[65]

September–1 October 1983, NASA TM-84920); Stephen D. Ishmael, Victoria A. Regenie, and Dale A. Mackall, "Design Implications From AFTI/F-16 Flight Test" (Ames Research Center, Moffett Field, CA: NASA TM-86026, 1984).

64. "NASA Facts: The X-29" (Dryden Flight Research Center, Edwards, CA: FS-2003-08-008-DFRC), available online at *http://www.nasa.gov/centers/dryden/pdf/120266main_FS-008-DFRC.pdf.*

65. Joseph R. Chambers, *Partners in Freedom: Contributions of the Langley Research Center to U.S. Military Aircraft of the 1990's* (Washington, DC: NASA SP-2000-4519, 2000), p. 110; Rumerman, *NASA Historical Data Book*, vol. 6, pp. 217–220; "NASA Facts: The X-29."

(NASA image EC89-0127-2)

Figure 5.3. X-29 aircraft no. 2 takes off from Edwards Air Force Base in 1989.

The first phase of flight testing covered design validation and performance, while the second phase examined high-angle-of-attack maneuverability. In the early 1990s, the Air Force fitted the second aircraft with jet nozzles in the nose in order to test an alternate method for controlling the aircraft at high angles of attack. In total, the two X-29s completed 422 research flights. At the most basic level, the X-29 proved that the enabling technologies, namely digital fly-by-wire and advanced composite construction, were up to the task. The design was controllable, and the wings retained their structural and aerodynamic integrity. This was an important validation of research that NASA had conducted into both technologies in the prior decade. Furthermore, flight testing showed that the X-29 could be flown at extreme angles of attack and remain controllable, just as had been hoped. It was fully controllable up to 45 degrees angle of attack (meaning that if the aircraft was moving in a horizontal direction, the nose could be pointed up at a 45-degree angle). It retained some controllability up to 67 degrees. One of the anticipated gains was a significant decrease in aerodynamic drag. The aircraft's wings were thin and employed a fine supercritical airfoil. Also, a dynamic stability system manipulated control surfaces such that both the canards and wing generated lift. Unfortunately, flight tests did not demonstrate the expected levels of drag reduction.[66]

---

66. "NASA Facts: The X-29."

F-18 High Alpha Research Vehicle (HARV)

NASA was not sitting on the sidelines through the X-29 program. Fighter aircraft design was of special interest to researchers at Langley, which had its own High Alpha Technology Program (HATP) office. Drawing on the strengths of all four aeronautics Centers, HATP engineers modified an existing fighter jet and flight-tested it from 1987 to September 1996. NASA's High Alpha Research Vehicle, or HARV, was stable up to 70 degrees angle of attack, and could be rolled at up to 65 degrees. This was a remarkable accomplishment, proving the utility of an innovative technique, thrust vectoring.[67]

Langley's HATP office opened in 1985, one year after the X-29 program began its own flight tests. Whereas the X-29 program relied on a radically new aircraft configuration, NASA's researchers chose to modify an existing aircraft. They acquired a preproduction McDonnell Douglas F-18, an aircraft that was already very capable in its own right, as it could fly to a 55-degree angle of attack. Dryden personnel rebuilt and instrumented the aircraft. The first phase of the program, which saw 101 flights from 1987 to 1989, was one of data collection. NASA researchers learned how to measure aerodynamic flows in this unusual flight regime. The second phase saw the addition of thrust vectoring, that is, six paddle-like devices (three for each of the aircraft's two engines) that could redirect exhaust. This was a highly complex modification requiring multidisciplinary expertise. As with the HiMAT and X-29, the F-18 required special flight control computers and software tailored to the aircraft's unique control mechanisms. The pilot would not have special controls for thrust vectoring; rather, the paddles needed to respond to the pilot's traditional inputs. Adding a new level of sophistication, the paddles allowed for three-dimensional thrust vectoring. They could be used to control yaw as well as pitch.[68]

Phase 2 testing ran from 1991 to 1993 and covered 193 flights. It was at this point that NASA's test pilots were achieving 70-degree angles of attack and the ability to roll at 65 degrees. By way of contrast, the unmodified F-18 would have been difficult to roll beyond a 35-degree angle of attack. The last test phase ran from 1995 to 1996 and saw the addition of movable nose strakes. Not unlike the jet nozzles used in the last set of X-29 tests, the nose strakes

---

67. "NASA Facts: F-18 High Angle-of-Attack (Alpha) Research Vehicle" (Dryden Flight Research Center, Edwards, CA: FS-2003-08-002-DFRC); Kenneth W. Iliff and Kon-Sheng Charles Wang, *Flight-Determined Subsonic Longitudinal Stability and Control Derivatives of the F-18 High Angle of Attack Research Vehicle (HARV) with Thrust Vectoring* (Edwards, CA: NASA TP-97-206539, 1997); Albion H. Bowers et al., *An Overview of the NASA F-18 High Alpha Research Vehicle* (Edwards, CA: NASA TM-4772, 1996).
68. "NASA Facts: F-18 High Angle-of-Attack (Alpha) Research Vehicle."

sought to control the vortices that flowed off the nose. The strakes themselves were hinged flaps that, when deployed, flipped into the airstream curling about the nose. Through 109 flights, test pilots examined different ways of employing the strakes alone and in conjunction with thrust vectoring. As before, the aircraft's control laws had to be well understood and programmed into the flight computer so that the operation of the various devices was automatic.[69]

X-31 Enhanced Fighter Maneuverability Demonstrator

The Defense Advanced Research Projects Agency, which had jointly sponsored the X-29, returned with one more experimental vehicle in the early 1990s. The X-31 Enhanced Fighter Maneuverability Demonstrator, or EFM, flew from 1992 to 1995. In addition to participation from the U.S. Air Force, U.S. Navy, and NASA, the Federal Republic of Germany joined as an international partner on this project. Rockwell and Deutsch Aerospace designed and constructed two prototype jet aircraft; their primary distinguishing feature was thrust vectoring using a three-paddle design similar to that of NASA's HARV. One new feature of these paddles was their construction: whereas the HARV used a high-temperature alloy, Inconel, the X-31 used "carbon-carbon," a specially designed carbon fiber that could withstand high-temperature jet engine exhaust. The aircraft also incorporated canards similar to those on the X-29.[70]

At Dryden, NASA test pilots and engineers worked alongside their counterparts from the military and from Germany. As the team expanded the envelope of operations (i.e., increased the angle of attack), they encountered various control difficulties, including yawing and lurching. With each problem, the team modified the aircraft and pushed forward. Eventually, they exhibited controlled flight to 70-degree angles of attack, including rolling the aircraft at the same angle, and showed that the aircraft could use this capability to outmaneuver conventional fighter aircraft in simulated battles.[71]

X-36 Tailless Fighter Agility Research Aircraft

The X-36, although a 1990s project, is appropriately placed in the line of experimental vehicles that began with the HiMAT. Sponsored by NASA's Office

69. Ibid.

70. Peter Huber, "X-31 High Angle of Attack Control System Performance," in *Fourth High Alpha Conference*, vol. 2, Dryden Flight Research Center, Edwards, CA, 12–14 July 1994, NASA CP-10143).

71. Dave Canter, "X-31 Post-Stall Envelope Expansion and Tactical Utility Testing," in *Fourth High Alpha Conference*, vol. 2, Dryden Flight Research Center, Edwards, CA, 12–14 July 1994, NASA CP-10143).

of Aero-Space Technology and McDonnell Douglas (purchased by Boeing during the program), the X-36 began as a collaboration between Ames and McDonnell Douglas. The X-36 was, like the HiMAT, a remotely piloted scale aircraft, but unlike the air-dropped HiMAT, the X-36 could take off under its own power. NASA flew the McDonnell Douglas–built aircraft 31 times in 1997, experimenting with and validating the tailless design. As with all the prior high-maneuverability vehicles, this was an unstable design that required digital flight control. By the mid-1990s, however, digital fly-by-wire was sufficiently mature that NASA and Boeing could make use of some commercially available components. This was a far cry from the time they invested in building the original digital fly-by-wire equipment in the 1970s. Interestingly, the Air Force Research Lab conducted a follow-on program that studied the use of adaptive controls, a kind of software that could reconfigure itself in order to compensate for failed or damaged controls. In essence, the software was creating new flight control laws on the fly.[72]

The National Aero-Space Plane and the Orient Express

The National Aero-Space Plane was an ungainly product of political circumstance and technological boosterism. NASP was an attempt to build a hypersonic, single-stage-to-orbit vehicle (a.k.a. the X-30), but for a brief period, it was also associated with an attempt to create a high-speed commercial transport called the Orient Express. Before the NASP program, the Department of Defense had been conducting research on hypersonic vehicles under a confidential project called "Copper Canyon." NASP, initiated in December 1985, represented phase 2 of this research. Parallel to the DOD's earlier research, NASA had its own interest in both supersonic and hypersonic research. There was a cohort of researchers who firmly believed in the viability of an American-designed and -built supersonic transport. Langley engineers also had conducted experiments and limited design studies of various hypersonic engines. By the late 1970s, John Becker, John Henry, and Robert Jones supported a scramjet design that integrated the fuselage and engine into a single package. Supersonic work was, for a period, conducted as part of the Supersonic Cruise Aircraft Research program. Funding was summarily ended and transferred to the human space program in 1981 in what appeared to be a preview of future

72. Mark Sumich and Rodney O. Bailey, "X-36 Tailless Fighter Agility Research Aircraft," in Research and Technology 1997 (Ames Research Center, Moffett Field, CA: NASA TM-98-112240), pp. 65–66.

cuts to federally funded civilian R&D. NASP helped revive NASA's work on supersonic and hypersonic vehicles.[73]

NASP also answered NASA's critics who maintained that the Agency lacked a long-range, risk-taking vision akin to that of the Apollo era. While this had been the modus operandi in the human space program, aeronautics had long been a research organization with no specific programmatic goals. True, aeronautics did carry out research that sought specific ends, such as specified gains in engine efficiency, but aeronautics (and the whole of OAST for that matter) was not supposed to be program oriented, at least not as envisioned at the beginning of NASA. By the 1980s, however, the critique of NASA's vision, or lack thereof, was taking greater hold. A committee that reported to the White House Office of Science and Technology Policy concluded similarly that NASA needed to reinvigorate itself. The committee recommended three radical goals, among them an air-breathing hypersonic space-plane.[74]

President Reagan's science advisor, George Keyworth, aware of both the DOD and NASA hypersonic programs, promoted the idea of a unified effort. The technology appeared feasible; it would give NASA's aeronautics a decisive vision and would solve the problem of limited access to space, one of the stumbling blocks in Reagan's Strategic Defense Initiative.[75] Reagan kicked off the NASP program in his 1986 State of the Union Address:

> And we are going forward with research on a new Orient Express that could, by the end of the next decade, take off from Dulles Airport, accelerate up to 25 times the speed of sound, attaining low Earth orbit or flying to Tokyo within 2 hours. And the same technology transforming our lives can solve the greatest problem of the 20th century. A security shield can one day render nuclear weapons obsolete and free mankind from the prison of nuclear terror. America met one historic challenge and went to the Moon.

---

73. Conway, *High-Speed Dreams*, pp. 187, 209; Andrew J. Butrica, *Single Stage to Orbit: Politics, Space Technology, and the Quest for Reusable Rocketry* (Baltimore: Johns Hopkins Press, 2003), pp. 78–82; Larry Schweikart, *The Hypersonic Revolution: Case Studies in the History of Hypersonic Technology*, vol. 3, *The Quest for the Orbital Jet: The National Aero-Space Plane Program (1983–1995)* (Washington, DC: Air Force History and Museums Program, 1998).
74. Conway, *High-Speed Dreams*, pp. 204–206.
75. Ibid., pp. 212–213; U.S. General Accounting Office (GAO), Report to Congressional Requesters, "National Aero-Space Plane: Restructuring Future Research and Development Efforts" (GAO/NSIAD-93-71, #148140, December 1992), pp. 13–15.

> Now America must meet another: to make our strategic defense real for all the citizens of planet Earth.[76]

The announcement involved a small sleight of hand: while Reagan did link NASP to the Strategic Defense Initiative, the vision of NASP was of a commercial vehicle. The appellation "Orient Express" successfully distracted viewers from the program's underlying military objectives.[77]

Just how serious the White House was about the commercial vision of the Orient Express is doubtful, and not just because this would have represented an ideologically repellent act of picking a winner. (According to free-market dogma, markets provide the most efficient mechanism for choosing among technologies.) The problem with the Orient Express was that the craft would have spent most of its time either accelerating or decelerating, an inefficient and, for passengers, uncomfortable proposition. Additionally, the vehicle's liquid hydrogen fuel would have required costly installations. NASP was a spacecraft; it entailed technologies and operations quite distant from what commercial airlines might consider feasible. As recounted in the following chapter, NASP did give a boost to Langley's supersonic commercial aircraft efforts, lending political support for an eventual High Speed Civil Transport and the High Speed Research program.[78]

DOD, of course, was undeterred by the lack of a commercial application of a hypersonic vehicle. From 1986 to 1993, DOD spent over $1.2 billion on the project. NASA spent nearly $400 million. The initial set of technologies that the NASP program envisioned were hydrogen-fueled supersonic ramjet engines (scramjets), new lightweight and heat-resistant materials, an integrated engine/airframe, and the use of slush hydrogen (ice/liquid mixture) for cooling and fuel. The program quickly ran into numerous technical obstacles. The major subcontractors determined that the original specifications were unfeasible and technically unsubstantiated.[79] Overall, the NASP program objectives also went through a number of changes, initially focusing on an experimental vehicle, then an operational vehicle with a large payload in 1988, and finally a research program in 1989. The U.S. General Accounting Office (GAO), in a 1992

76. "Address Before a Joint Session of Congress on the State of the Union, 4 February 1986," Personal Papers, file SP230-86, Ronald Reagan Presidential Library, National Archives and Records Administration.
77. See Butrica, *Single Stage to Orbit*, chap. 4.
78. Conway, *High-Speed Dreams*, pp. 217–219.
79. U.S. GAO, "National Aero-Space Plane," p. 11; Butrica, *Single Stage to Orbit*, p. 79.

Table 5.1. NASP Appropriations, 1986 to 1993, Dollars in Millions

	1986	1987	1988	1989	1990	1991	1992	1993	Total
DOD	45	110	183	231	194	163	200	150	1,276
NASA	16	62	71	89	60	95	5	0	398
Total	*61*	*172*	*254*	*320*	*254*	*258*	*205*	*150*	*1,674*

Source: U.S. General Accounting Office, Report to Congressional Requesters, "National Aero-Space Plane: Restructuring Future Research and Development Efforts" (GAO/NSIAD-93-71, #148140, December 1992), table 1.1, p. 15.

report to Congress on NASP, noted, "The NASP Program's 7-year history has been characterized by turmoil, changes in focus, and unmet expectations."[80]

Though NASP funding ceased in the early 1990s, the idea of a single-stage-to-orbit vehicle continued elsewhere in the form of the DCX and the X-33. No one confused these latter vehicles with commercial jet transportation.

Icing and Wind Shear

Perhaps the farthest one could get from the big-budget, high-risk, top-down approach of NASP was NASA's work on icing and wind shear. These were classic nuts-and-bolts research projects that successfully addressed pressing aviation concerns. Like NASA's energy-efficiency and noise-reduction programs of the 1970s, icing and wind-shear research were emblematic of the Agency's responsiveness to public outcries in the 1980s. The Agency had long addressed safety problems, and these particular dangers came to the fore through highly publicized tragedies.

The first accident that grabbed the public's attention was an Air Florida 737 taking off from National Airport in January 1982. Wing and engine icing caused the aircraft to lose lift and control on climb-out; the aircraft plunged into the Potomac River and the 14th Street Bridge. Of the 79 persons on board, 74 died. Television footage of rescuers braving the icy waters of the Potomac riveted the nation. Subsequent accidents reinforced the danger of icing: in 1985, a DC-8 carrying U.S. service members crashed on takeoff, taking the lives of 256 people; and in 1987, a DC-9 crashed on takeoff, killing all 28 on board. Both aircraft lost lift and controllability due to icing.[81]

Icing research at the NACA dated to 1928 at Langley. The first icing wind tunnel was a refrigerated device with a paltry 6-inch cross section. Icing

80. U.S. GAO, "National Aero-Space Plane," pp. 4, 16–17.

81. William M. Leary, *"We Freeze to Please": A History of NASA's Icing Research Tunnel and the Quest for Flight Safety* (Washington, DC: NASA SP-4226, 2002), pp. 81–82, 107.

research moved from Langley to the new Aircraft Engine Research Laboratory, or Lewis Lab, during World War II. There, a much larger 6-by-9-foot refrigerated tunnel would be the centerpiece of the NACA's icing research for the coming decades.[82] In the early years of NASA, the Agency did not put much emphasis on icing research, but because of continued interest from industry, the Lewis Icing Research Tunnel (IRT) remained in operation. By the 1980s, the old IRT was feeling its age, and so from 1986 to 1988, NASA closed the tunnel for renovation.[83]

By the time the tunnel reopened, it was available to work on the causes behind the ice-related commercial aircraft crashes. Most of these tragedies had in common the use of deicing fluids. Boeing and the Association of European Airlines (AEA) had been examining the effectiveness of these agents for a number of years; when the IRT reopened in 1988, it also began to evaluate the fluids. Researchers needed to understand how the fluids worked at different temperatures and under different weather conditions, as well as how long the fluids lasted before they needed to be reapplied. Finally, in November 1992, the FAA was able to issue new deicing guidelines based on the body of research conducted by Boeing, the AEA, and NASA. Concurrently, the Lewis Lab sponsored the development of computer codes to mathematically model ice formation. Developed initially at the University of Dayton by Professor James K. Luers, the codes (eventually dubbed LEWICE) were refined and validated by Lewis researchers. LEWICE, like the computational fluid dynamics codes developed elsewhere at NASA in the 1970s and 1980s, became an important and very cost-effective tool for industry.[84]

Public concern over wind shear began with the August 1985 crash of a Delta L-1011 on approach to landing at Dallas–Ft. Worth airport. A highly localized weather system shifted wind directions so violently that the wide-body jet lost airspeed and crashed. Of the 163 persons on board, 137 perished. Of course, wind shear was not a new natural phenomenon. It had periodically claimed aircraft and lives. What was unusual was the public's awareness. As with icing, there had been prior work on wind-shear detection. In the 1970s, the FAA studied the problem and installed the Low Level Windshear Alert System (LLWAS) at major airports. LLWAS compared wind velocity at different points on the ground around the airport. Significant differences triggered a wind-shear alert. Unfortunately for the Delta crash, the LLWAS system installed at

82. Ibid., pp. 2–6, 19–22.

83. Ibid., pp. 97–101.

84. Ibid., pp. 105–119.

Dallas–Ft. Worth gave its warning only after the crash.[85] Popular and political concern led to the creation of the National Integrated Wind Shear Plan in 1986. The FAA-managed program sought to develop new radar systems that could detect wind shear (namely, the Doppler weather radar), an airborne detection system, and improved communication and training. The airborne system was NASA's contribution to the program.[86]

NASA Langley oversaw the wind-shear research using the 737 airborne laboratory, the Transport Systems Research Vehicle (TSRV). This was a logical platform, for as a commercial-type aircraft, it would bring NASA closer to approximating actual conditions. It also would help shorten the development of commercially available applications. Furthermore, the TSRV was an appropriate base for the kind of systems engineering that stretched from manufacturers to NASA to the FAA. The engineers and pilots who worked with the TSRV understood that they were finding solutions in the integration of multiple expert groups and technologies. For wind shear, Langley examined three different types of forward-looking detection systems: microwave Doppler radar, Doppler light detecting and ranging (LIDAR), and passive infrared radiometry. The first two measured the Doppler shift of energy reflected off water particles; the third looked for temperature shifts that might signal a microburst of bad weather.[87] Langley researchers refined the systems and worked with the manufacturers to make them more reliable and integrated into cockpit and air traffic control procedures. In the summers of 1991 and 1992, Langley crews flew the detector-equipped TSRV through actual microbursts detected by ground-based Doppler radar. In the end, both Doppler radar and LIDAR were able to detect microbursts early enough to allow pilots to take action. The radar systems were more effective, however, and within a fairly short time, avionics manufacturers had equipment ready for commercial operations.[88]

Conclusion

From a purely technical perspective, the lesson that stands out from this decade of research is the degree to which a small number of revolutionary technologies enabled a whole set of new applications. Digital computers, composite materials, and advanced aerodynamics were making possible aircraft such as the X-29. Forward-swept wings had been tried before, indeed, long before, but now

85. Lane E. Wallace, *Airborne Trailblazer: Two Decades with NASA Langley's 737 Flying Laboratory* (Washington, DC: NASA, 1994), pp. 58–59.
86. Ibid., pp. 60–61.
87. Ibid., pp. 63–64.
88. Ibid., pp. 65–73.

could be flown successfully. Thus, NASA's fundamental research in the 1960s and 1970s helped set in motion an entirely new set of capabilities. Cold War funding sped the development of highly unstable yet maneuverable aircraft, but this was a technical and tactical trend that was already well under way. A fourth fundamental technology, stealth, was arguably the most revolutionary of the aeronautical developments in the 1980s, but that narrative will have to wait for a day when historians have open access to those records.

NASP was illustrative of the kind of ideological battle taking place in Washington, far away from the labs. A White House that, in its heart, did not believe in big government was initiating a large-scale technology development program. NASP drew support from those who believed in supersonic and hypersonic flight, as well as wonks who believed that NASA's aeronautics program, like the space program, suffered from a lack of Apollo-like goals. Cobbled together without the kind of technological underpinnings that made the ACEE program a nearly automatic success in the 1970s and without the kind of money that would be required to see it through, NASP was truly a high-risk endeavor, politically and technically. NASP did fund high-speed research and so was not necessarily a net loss from the researchers' standpoint, but it came at the cost of increased politicization and a shift toward more top-down research management.

Chapter 6:

The Icarus Decade

Before funding came crashing down at the close of the 1990s, aeronautics would enjoy one last upswing. Its budget would approach $1 billion in 1994; it would be running two of its largest programs ever, the Advanced Subsonic Technology (AST) Program, which peaked at $173 million in 1997, and the High Speed Research (HSR) Program, which reached its zenith the following year at nearly a quarter of a billion dollars. By 2000, both programs would no longer exist and the aeronautics budget would fall to just $600 million. Adjusted for inflation, this was a 45 percent drop in funding over seven years. At first glance, there are obvious geopolitical and industrial shifts that presaged aeronautics' difficulties, but this was more than just the same agency caught in difficult circumstances. Aeronautics had, with AST and HSR, placed the bulk of its widely dispersed research into two baskets. As a funding strategy and as an organizing principle, it was unsustainable. Even as aeronautics' budget was on the rise, so were expenditures for the International Space Station and Space Shuttle. In the competition for program dollars, AST and HSR would not win out over the space program. In the end, the mid-decade surge in aeronautics funding belied increasingly fragile political and Agency support.

The end of the Cold War was the most fundamental shift in the history of NASA's aeronautics program. Viewed from the present, it may be difficult to grasp the importance of the USSR's aeronautics and military programs as an overwhelming justification for NASA's own research. However, when it came to aeronautical R&D, questions about the proper federal role in R&D or commercial utility were trumped by the more primal goal of ensuring an unassailable technological advantage over the Soviet Union. Not all of NASA's aeronautics work was military-related, of course, but many areas had wide applicability (e.g., basic wind tunnel research) and so benefited from this logic. The Soviet Union's collapse brought discussion of a peace dividend through a reduction in security-related expenditures. Directly, NASA had to explain why aeronautics research was still essential despite a clear and growing technological gap between the United States and any potential enemy on the horizon. Indirectly, NASA's aeronautics programs were beginning to see fewer military development programs. Without a security imperative, justifications for NASA fell more heavily on commerce, thus inviting questions about the appropriate role of the federal government in commercially oriented R&D. The drop in military programs also had a colossal impact on NASA's industrial partners, triggering a wave of consolidation that left the field with half as many players as in the 1980s and, in like fashion, reducing the breadth of lobbying on NASA's behalf.[1]

Even as the pipeline for new fighters and bombers slowed to a trickle, the environment for nurturing commercial development became more complicated. In the area of large commercial aircraft, Airbus was no longer a government-subsidized anomaly feeding off the scraps of Boeing, Lockheed, and McDonnell Douglas. Airbus was gaining significant market share and finally had a product that was selling briskly. Its A320 narrow-body aircraft with an efficient wing and digital fly-by-wire controls put U.S. airframe manufacturers on the defensive. From 1979 to 1989, the American share of the world civilian aerospace market had declined from 76 percent to 58 percent, while the European share had risen from 22 percent to 37 percent.[2] By 1992, American concern over Airbus and the launch aid it received from European member companies led to a bilateral accord between the European Union (EU) and the United States. European governments were limited to financing 33 percent of development costs, aid that was to be repaid with interest if the aircraft made money. The Europeans, for their part, demanded and received an upper limit to the amount of R&D aid that could come from the U.S. government (set at 3 percent of the industry's large commercial aircraft turnover).[3] Thus, at a time when American aircraft manufacturers were threatened by foreign competition on the one hand and reduced military expenditures on the other, the United States could not appear hypocritical about corporate subsidy. NASA's commercially oriented R&D ran the risk of undermining the United States' opposition to European launch

1. Lockheed purchased General Dynamics' military aircraft division in 1993. In 1994, Northrop purchased Grumman Aerospace. Boeing purchased Rockwell's space and defense businesses (including North American Aviation) in 1996. As noted below, Boeing merged with McDonnell Douglas in 1997.
2. U.S. General Accounting Office, Report to the Honorable Lloyd Bentsen, U.S. Senate, "High-Technology Competitiveness: Trends in U.S. and Foreign Performance" (GAO/NSIAD-92-236, GAO #147990, September 1992), table I.4, p. 53.
3. John Newhouse, *Boeing Versus Airbus: The Inside Story of the Greatest International Competition in Business* (New York: Knopf, 2007), pp. 46–47. For a very helpful study of the 1992 bilateral agreement, see U.S. General Accounting Office, "International Trade, Long-Term Viability of U.S.–European Union Aircraft Agreement Uncertain" (GAO/GGD-95-45, December 1994). Concern over the rise of the European aeronautics sector (principally Airbus) led to some amount of hand-wringing over Europe's aeronautical R&D. See U.S. General Accounting Office, Report to Congressional Requesters, "European Aeronautics: Strong Government Presence in Industry Structure and Research and Development Support" (GAO/NSIAD-94-71, GAO #151432, March 1994).

aid. Furthermore, as the commercial aircraft industry consolidated (Lockheed did not produce another commercial aircraft after the L-1011, and Boeing purchased McDonnell Douglas in 1997), there remained only one domestic beneficiary of NASA's research on large commercial aircraft. Short of limiting work to safety issues, propulsion, small aircraft, military research, and air traffic control, doing something for the good of the industry really meant doing something for the good of Boeing. Although the 1992 bilateral accord held until the next decade, there could be little question that any attack on EU subsidies would invite a similar scrutiny of the U.S. government's various R&D programs. This is exactly what came to pass in the following decade when, as Airbus attained a majority of the large commercial aircraft market, President George W. Bush's administration chose to press the issue.[4]

In spite of what were sea changes to the macro-political and economic environment for aeronautics, the impact on NASA initially was muted by a favorable attitude toward federal R&D (and science funding in general) under President William Jefferson Clinton, as well as a robust economy for part of the decade. Indeed, the Clinton White House specifically saw federal R&D as a necessary input to the health of the private sector.[5] With regard to support for the country's commercial aircraft manufacturers, Clinton was initially very

4. The U.S. filed its dispute (a procedure that begins with a "request for consultations") with the World Trade Organization on 6 October 2004 (WTO #DS316). The EU countered the same day (WTO #DS317). The Europeans' complaint stipulated the following NASA programs as prohibited subsidies under the 1994 General Agreement on Tariffs and Trade: High Speed Research Program, Advanced Subsonic Technology Program, Aviation Safety Program, Quiet Aircraft Technology Program, High Performance Computing and Communications Program, Research & Technology Base Program, Patent Waiver Program, any NASA personnel working on Boeing R&D, and NASA procurement contracts. World Trade Organization, "United States—Measures Affecting Trade in Large Civil Aircraft, Request for Consultations by the European Communities," 12 October 2004, Document # 04-4275, WT/DS317/1, G/L/698, G/SCM/D63/1. Supporting the argument that U.S. industry was a special beneficiary of NASA's aeronautics research was the Agency's policy of restricting access to information deemed of commercial and/or proprietary value (through "limited distribution" or "early domestic distribution"). Such restrictions, however, have been applied to only a very small fraction of NASA's research output. See U.S. General Accounting Office, Report to Congressional Requesters, "NASA Aeronautics: Protecting Sensitive Technology" (GAO/NSIAD-93-201, GAO 149782, August 1993), pp. 7–8.
5. On the Clinton administration's attitude, see W. D. Kay, *Defining NASA: The Historical Debate over the Agency's Mission* (Albany: SUNY Press, 2005), pp. 161–162; and Edmund L. Andrews, "Clinton's Technology Plan Would Redirect Billions from Military Research," *New York Times* (24 February 1993): Technology section.

critical of the bilateral agreement and argued that the United States needed to match Europe's industrial support. After meeting with industry representatives, the President chose to overlook the issue of subsidies. NASA's aeronautics funding reached its highest point the following year (figure 6.1).[6]

Another of Clinton's early decisions, however, had long-term ramifications for the aeronautics budget: continuing to build the Space Station. Survival of the Space Station, a program that dated to the Reagan presidency, was hardly assured when Clinton assumed office. Clinton did agree to a leaner, "cost-effective design" and, in his budget proposal, gave NASA $2.1 billion of the $2.3 billion that it requested for the Station.[7] The Space Station would, not surprisingly, suffer cost overruns, a trajectory that put it on a collision course with aeronautics' growing appetite.

Against the broad trends in politics and markets in the 1990s, changes were also afoot in how NASA operated. Dan Goldin, appointed to the post of Administrator by George H. W. Bush in 1992, sought to make NASA more efficient and businesslike. Goldin's own career had taken him through NASA, where he was once an engineer at Lewis in the early 1960s, to TRW Aeronautical Systems, where he was vice president and general manager of the Space Technology Group. "Faster, better, cheaper," became the Agency's unofficial (and borrowed) motto, and though Goldin's primary concern was space, the organizational changes were widespread.[8] Among Goldin's reforms were a greater reliance on subcontracting to the private sector, implementation of strategic management methodologies, and a preference for what were perceived as lower-cost (yet technically advanced) technologies.[9] Goldin, specifically chosen by Vice President Dan Quayle to shake things up, came to symbolize drastic change at NASA, although his key efforts sometimes reflected wider movements within the federal government. For example, both the Clinton White House and Republican fiscal conservatives championed a more responsive,

6. Gwen Ifill, "Clinton to Fight Foreign Subsidies," *New York Times* (23 February 1993): section A, p. 1; Newhouse, *Boeing Versus Airbus*, pp. 47–50. Newhouse mischaracterizes Clinton's plan to spend $8 billion in federal funding for aeronautics R&D over the coming 5 years. In fact, the money was for more than just aeronautics. Nevertheless, Clinton was prepared to support great investment in aeronautics R&D. Executive Office of the President, Office of the Press Secretary, "Remarks by the President to Employees of Boeing," press release, 22 February 1993.
7. William J. Broad, "Clinton's Economic Plan: The Space Program: Space Station, Trimming Back, Survives the Ax," *New York Times* (19 February 1993): section A, p. 17.
8. "Faster, better, cheaper" had its roots in the Strategic Defense Initiative Organization, where it was "faster, cheaper, smaller." See Andrew Butrica, *Single Stage to Orbit*, pp. 60, 149–151.
9. Kay, *Defining NASA*, pp. 158–159.

Figure 6.1. Aeronautics Program Budget, 1990–2000.

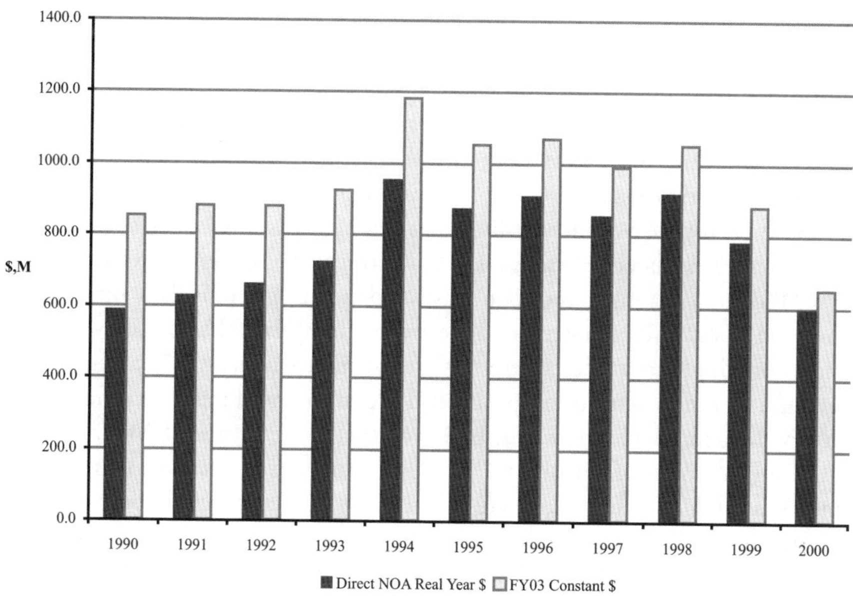

(Figures provided by the Office of Aerospace Technology, NASA.)

slimmed-down bureaucracy with emphasis on increased public-private partnerships and modern business methods, such as total quality control.[10] The problem of the faster, better, cheaper approach was that while this might have been an entirely appropriate response to large, bureaucratic space programs, the aeronautics branch had a long history of being smaller and cheaper (though not necessarily faster). Beyond NASP, aeronautics did not have the budgetary wherewithal to create large, ongoing programs. Ironically, aeronautics programs in the Goldin era became larger and more expensive.

Goldin's "strategic management" was an organizational framework for establishing long-term priorities (and consequently budgetary goals), as well as benchmarks for evaluating progress. The Agency reorganized itself around the concept of "Enterprise Management," i.e., Headquarters and four to five Enterprise groups (depending on the timeframe): Mission to Planet Earth, Aeronautics, Human Exploration & Development of Space, and Space Science.[11] Consistent with the organizational evolution that had been in the works over the prior decades, the strategic management reorganization further centralized research administration. The NASA Administrator was now

10. Butrica, *Single Stage to Orbit*, pp. 150–151, 172–174.

11. NASA, Strategic Management Handbook, October 1996.

ensconced within a planning apparatus, surrounded by top-level planning committees with soft organizational links to the Centers (e.g., Center Directors served on the Senior Management Council, which was an advisory body to the Executive Council and Administrator). At the Enterprise Management level, the Enterprise Associate Administrators were the "customer interface" for NASA. The Enterprise areas were NASA's "primary business areas for implementing NASA's mission and serving customers."[12]

Piled on top of Goldin's business-minded reforms was a dose of new congressionally mandated oversight. The Government Performance Results Act of 1993 (GPRA), seeking to root out "waste and inefficiency in Federal programs," imposed a new system of performance accounting.[13] For NASA, GPRA meant that research programs had to establish performance benchmarks at the outset (extending a minimum of five years into the future) and report on the degree to which such benchmarks had been achieved. The Office of Management and Budget would administer GPRA compliance. In the context of scientific and technological research, where subjects are studied precisely because they are unknown, GPRA perversely asked NASA to specify and meet outcomes that could not be predicted. Truly, the contrast with the early days of NASA could not be greater. Where branch managers and Center Directors once evaluated the performance of their in-house research programs, OMB and NASA Headquarters now evaluated research efficacy by using data from "objective, quantifiable, and measurable" goals. It not only added to the costs of research administration at Headquarters, but it required a significant amount of up-front planning and continued reporting paperwork on the part of researchers.

Complicating the picture of NASA policy-making, President Clinton added a new voice when he established the National Science and Technology Council (NSTC) as part of the President's Office of Science and Technology Policy in 1993. While the NSTC may have had the effect of burying space policy deeper within the White House administration, it created another forum for discussing aeronautics R&D policy.[14] In 1994, the NSTC created a Committee on Transportation R&D.[15] This committee resembled the original NACA main

12. Ibid, sec. 3.3.1.
13. Government Performance and Results Act of 1993, Pub. L. 103-62 (1993). See also NASA Strategic Plan, Aeronautics Enterprise, February 1995.
14. W. D. Kay, *Defining NASA*, p. 161.
15. White House Office of Science and Technology Policy, National Science and Technology Council, Committee on Transportation Research and Development, Intermodal Transportation Science and Technology Strategy Team, "Transportation Science and Technology Strategy," September 1997, p. iii. The committee, or subcommittee more properly, was composed of members of

committee established in 1915 in that it created a small, high-level committee to advise the executive branch. It differed in that it was not restricted to aviation, it was not heavily weighted by defense interests, and it did not have representation from the private sector. Still, it gave NASA's aeronautics policy the kind of standing high-level platform that it had lacked since at least the institution of NASA. The Committee would issue its first set of recommendations in 1997 ("Transportation Science and Technology Strategy"), and two more reports in 1999 ("Transportation Strategic Research Plan" and "Transportation Science and Technology Strategy"). Aside from the specific recommendations, what is distinctive is the emphasis placed on "users, industry, and other stakeholders" in defining national goals, as well as the use of "partnership initiatives" to pursue R&D.[16] Users, industry, and stakeholders were to identify the most pressing projects, and the programs were to be implemented as public-private initiatives that would guide and carry out the R&D. The NSTC's recommendations fit squarely with the Administration's broader goal of "reinventing government."

Aeronautics' Big Programs

The most unusual feature of the aeronautics programs in the 1990s was the move to concentrate most of the nonbase funding into two major programs: AST and HSR.

Packaged programs, like AST and HSR, were not new. The Aircraft Energy Efficiency Program (ACEE) had, in the 1970s, ushered in the strategy of cobbling smaller programs together in order to address pressing national needs. In looking at separate programs, Congress might not have recognized that NASA was, in fact, doing something about energy efficiency, but AST and HSR took the idea much further, becoming umbrella programs for nearly everything aeronautics related. As table 6.1 shows, by 1996, AST, HSR, and the Numerical Aerodynamic Simulation program were the only three major

the Department of Commerce (1), the Department of Transportation (4, including the Chair), OMB (1), the EPA (1), the Department of Energy (1), NSF (1), the Department of Defense (1), NASA (1, the Vice-Chair), and OSTP (1, White House Co-Chair). See also National Science and Technology Council, Committee on Technology, Subcommittee on Transportation R&D, "National Transportation Science and Technology Strategy," April 1999.

16. The mandate of the subcommittee was to "ensure that Federal investment in transportation R&D is (1) coordinated to ensure efficient use of Federal funds aimed at this mission; (2) focused on projects identified by users, industry, and other stakeholders as being the most critical to achieving success in agencies' missions; and (3) limited to areas where it is clear that major public benefits can only be achieved through cost-shared Federal research." OSTP "Transportation Science and Technology Strategy," p. iii.

Table 6.1. Primary Aeronautics Budgets at NASA, 1990–2000 (millions of dollars)

	1990	1991	1992	1993	1994	1995	1996	1997	1998	1999	2000
AST (Advanced Subsonic Technology)			5.0	12.4	89.3	125.8	169.8	173.6	144.4	89.6	
HSR (High-Speed Research)	24.5	44.0	76.4	117.0	197.2	221.3	233.3	243.1	245.0	180.7	
NAS (Numerical Aerodynamic Simulation)	41.8	44.1	45.4	47.9	48.1	46.2	48.1	38.6			
Rotorcraft	3.6	5.1	4.9	7.0							
HPAST (High Performance Aircraft Systems Technology)	9.7	10.5	10.7	12.1							
AdvProp (Advanced Propulsion System Technology)	13.1	15.0	15.2	16.9							
MatStruct (Materials and Structural Systems Technology)	28.1	39.9	37.5	36.6	25.7	24.3					
AeroSP (Aerospace Plane Technology)	59.0	95.0	4.1		20.0						
HypSonic (Hypersonic Program)				33.1	26.0						
AvSafety (Aviation Safety)											64.3
UEET (Ultra-Efficient Engine Technology)											68.3
AvSyCap (Aviation Systems Capacity)									56.7	53.9	62.9
QuietAc (Quiet Aircraft Technology)											18.3
R&TBase (Research and Technology Base)	321.8	336.4	343.3	451.5	420.3	354.3	430.6	404.2	428.3	424.1	405.8

Source: Figures provided by the Office of Aerospace Technology, NASA.

programs beyond base funding. Six other major programs had their funding eliminated, although what actually happened in many cases is that HSR and AST (primarily the latter) picked up elements of the old programs. Rotorcraft, for example, gained funding for civil tilt-rotors from AST's short-haul aircraft program. This new gambit also differed from the ACEE in that it did not address a pressing national problem, such as an energy shortage, and did not have easily identifiable end products. To be sure, the two programs did have concrete goals that they sought to meet, such as a 30-decibel reduction in perceived commercial aircraft noise, but as large packages, the collective impact did not present a unified image. Rather, the effort was diffuse, and as AST accrued new programs along the way and as HSR's timeline began to stretch out into the next millennium, these appeared to have no certain end point.

Advanced Subsonic Technology

The Advanced Subsonic Technology program had a gestation period of nearly a decade before NASA and OMB decided to fund it. Since 1983, NASA's aeronautics leadership had repeatedly asked for an expanded subsonic research program aimed at advancing the state of commercial aircraft. On the face of it, this should have been an easy sell; the subsonic category was the entire commercial market. If NASA was to help maintain the country's technological leadership, surely this segment demanded significant attention. Conspiring against the subsonic program were a few factors. The 1980s budget environment was particularly hostile to precisely this kind of federally funded research. Fiscal conservatives saw this as research that private industry should be conducting on its own and with its own funds. Others believed that NASA researchers needed to take greater risks, to address technological areas that the market had difficulty addressing on its own (e.g., supersonic research). Subsonic research, especially approaches that sought incremental improvements to the state of the art, had little revolutionary appeal. To put it more simply, AST was not the kind of research that fired the imagination. In 1990, the Aeronautics Directorate put together a $642 million program that stretched over five years, only to see it reduced by Headquarters and OMB to $120 million with only $17 million actually being allocated.[17]

The major factor behind AST's acceptance was the Clinton administration's and President Clinton's own desire to see NASA do more for the commercial

17. United States General Accounting Office, "NASA Aeronautics: Efforts To Preserve U.S. Leadership in the Aeronautics Industry Are Limited," testimony of Mark E. Gebicke before the Subcommittee on Government Activities and Transportation, Committee on Government Operations, House of Representatives, 18 March 1992.

aircraft industry. Clinton was mindful of the jobs that were being lost as McDonnell Douglas ceded more and more market share to Airbus. As old as AST was, it seemed tailor-made for the Clinton administration.[18] AST's ramp-up began slowly with $5 million budgeted in 1992 for aging aircraft research and fly-by-light/power-by-wire. The latter was an innovative approach to controlling aircraft systems with fiber-optic signaling and using electrical power in lieu of hydraulic systems.[19] The same two programs continued into 1993. The sea change came in 1994 with the introduction of eight new areas: noise reduction, terminal area productivity, integrated wing design, propulsion, short-haul aircraft, technology integration and environmental impact, environmental research aircraft and remote sensor technology, and composites. In 1999, the program was restructured into five areas: safety, environment, economics, reduced seat cost, and aviation systems capacity. By 2000, AST had been, in name and in much of the research it funded, eliminated.[20] The breadth of AST research precludes a comprehensive investigation here; two programs, composite materials research and noise research, will illustrate the character of the AST program.

Composites

Composite material research had long been a part of materials research at Langley. Composites were, as noted earlier, part of the ACEE program of the 1970s; work continued under the Advance Composite Technology (ACT) Program in the late 1980s. Prior to being incorporated into AST in 1996, the ACT program was funded to a level of between $10 and $30 million per year starting in 1989. Since the 1970s, composites had made large inroads into aircraft designs, their strength and lightness well proven by the early 1990s. One of the major impediments to more extensive use, however, was the high cost

18. In addition to Newhouse, *Boeing Versus Airbus*, pp. 46–47, see Douglas L. Dwoyer, telephone interview by Robert Ferguson, 16 June 2007, NASA HRC.
19. Power-by-wire was to see its first commercial application in the Boeing 787, which, as of this writing, had not yet flown. NASA conducted its research at Lewis with flight tests at both Dryden and Langley. Gale R. Sundberg, "Civil Air Transport: A Fresh Look at Power-by-Wire and Fly-by-Light" (prepared for the National Aerospace and Electronics Conference, Dayton, OH, 21–25 May 1990, NASA TM-102574); Anthony S. Coleman and Irving G. Hansen, "The Development of a Highly Reliable Power Management and Distribution System for Civil Transport Aircraft" (prepared for the 29th Intersociety Energy Conversion Engineering Conference, Monterey, CA, 8–12 August 1994, NASA TM-106697, AIAA-94-4107).
20. See NASA Aeronautics annual budgets, 1990 to 2000, in historical materials for *NASA's First A*, NASA HRC.

of manufacture. Reducing these costs through innovative fabrication methods was the prime focus of the ACT and AST composites study. NASA and its industrial partner, McDonnell Douglas (Boeing after the 1997 merger), chose to investigate a stitched, resin film infusion (S/RFI) composite wing. Instead of mechanical fasteners, such as rivets, the wing would be produced by stitching dry pieces of carbon fiber together, infusing the assembly with resin, and hardening the entire structure in an autoclave. S/RFI was ambitious, as was the decision to build an entire wing. The goal of the program was to create a wing that was 25 percent lighter, 20 percent less expensive to produce, and 4 percent less expensive to maintain.[21]

Langley oversaw the project, a large portion of which was subcontracted to McDonnell Douglas and carried out at company facilities in Long Beach and St. Louis (what Boeing would eventually call the Phantom Works Divisions). Structural testing would take place at Langley's labs. The wing that Langley and the Phantom Works engineers chose was a modified MD-90 wing (the MD-90 was a derivative of the Douglas DC-9). This saved them the trouble of designing an entirely new wing, but, more importantly, gave them a set of performance/cost benchmarks for judging the advantages of the S/RFI process. Phantom Works made some changes in the wing to accommodate manufacturing limitations but also took advantage of the new material's greater stiffness, creating a more slender and higher performance wing.[22]

The fabrication of the wing involved a large, specially designed and built, computer-controlled stitching machine. This was the third purpose-built stitching machine procured by NASA for the ACT program. Known as the Advanced Stitching Machine (ASM), it was a joint project of the Ingersoll Milling Machine Company and Pathe Technologies. ASM was an evolved design and incorporated lessons learned about needle design, sewing thread construction, and the handling of preformed materials.[23]

The Phantom Works team built a semispan composite test section and delivered it to Langley for structural analysis.[24] The team also built scores of "Design Development Test Articles," much smaller subassemblies that could be individually tested for various structural performance parameters (e.g.,

21. Michael Karal, *AST Composite Wing Program—Executive Summary* (Langley Research Center, Hampton, VA: NASA CR-2001-210650, March 2001), p. 1.
22. The composite wing had an aspect ratio of 12.1, whereas aluminum wings typically have an 8.5 aspect ratio. Karal, *AST Composite Wing Program*, pp. 1, 3, 21.
23. Ibid., pp. 61–64.
24. Ibid., p. 24.

strength, durability, or damage tolerance).[25] The team performed tests on the materials to understand how the resin flowed, how the fabric became saturated, and how it all cured under different production conditions. Additionally, the team evaluated the amount of savings in terms of reduced fasteners (an indirect measure of actual fabrication costs).[26] The finished product actually achieved a weight reduction of over 29 percent.[27] Extrapolating from learning curves, the program estimated a cost savings of 19.6 percent on the S/RFI wing box.[28] Structurally, the wing passed all of its tests and made it to 97 percent of its 2.5-g "design ultimate load" in the last, wing-up-bending test to failure.[29] From 1989 to 1997, expenditures on composite research, of which the ACT program was the primary component, topped $200 million.[30] There was consideration of a follow-on composite fuselage manufacturing program, but Boeing ultimately performed this research in-house with its own funds.[31]

Noise Reduction

Like the composite research, the AST noise program was part of a long line of noise-reduction research stretching back to the 1960s. Unlike the stitched-wing research, the noise program employed a shotgun approach to finding potential solutions. The program's goal was to achieve a 30-EPNdB (effective perceived noise level in decibels) reduction relative to the state of the art in 1992. Because no single technology would achieve this, the results would have to be pieced together through an assortment of technical and operational tricks. NASA conducted the program jointly with the FAA and partnered with four engine manufacturers: General Electric and Pratt & Whitney for large engines and Honeywell and Rolls-Royce for smaller engines. Langley oversaw the program and performed airframe analyses and scale testing of noise-reduction devices. The Glenn Research Center, formerly Lewis, conducted engine noise-reduction tests and furthered computational methods

25. Ibid., pp. 33–50.
26. Ibid., pp. 51–60.
27. Ibid., p. 19.
28. Ibid., p. 77.
29. Ibid., p. 83.
30. ACT expenditures were approximately $188 million, though in the final year of funding, costs were folded into the larger composites category.
31. Dwoyer telephone interview, 16 June 2007.

for predicting noise impact. Ames focused on computer analyses of airframe noise and carried out scale tests of devices.[32]

The program sought noise reduction in four areas: engines, nacelles and liners, the airframe, and community measures. Approximately 75 percent of the AST noise-reduction resources were focused on propulsion, the principal source of noise. Propulsion noise has evolved over the years with the growth of high-bypass turbojet engines. With these engines, most of the airflow comes from the large, slow-moving fan blades rather than the turbine core. Jet engines have grown quieter simply because the bypass air provides a smoother exhaust transition for the very turbulent gases of the engine core. As noise from the core has been reduced, noise from the fan has increased. Low-bypass engines have a distinct low-frequency rumble or roar, while high-bypass engines have higher-pitched whines. The remaining noise arises from the airframe, especially wing flaps and landing gear, as they are extended into the airstream during takeoff and landing.[33]

For airframe noise, researchers studied porous flaps and a cove filler for the leading-edge slat.[34] For engine noise suppression, they examined the use of scarf inlets, in which the lower portion of the inlet juts forward to bounce noise skyward, and better nacelle liners. Herschel-Quincke tubes were used to create tones that would passively suppress engine noise.[35] They tested the use of active noise control (creating out-of-phase sound waves), a technique they abandoned partway through the program. They looked at using different shapes on the engine stators (the nonmoving blades within the engine) and fan blades. They examined chevrons on the engine nozzles, where the air exhausts and mixes with outside air.[36]

Overall, the program had mixed success in meeting noise targets for all the engine/airframe combinations. Of the five configurations, one exceeded the program goal (30 EPNdB); two met the minimum goal (21 EPNdB); one fell just shy of the minimum; and one, the smallest engine, fell well short. As had been the case in noise reduction dating to the 1970s, the largest gains came simply from adopting the latest engines, which had higher bypass ratios.

---

32. Robert A. Golub, John W. Rawls, Jr., and James W. Russell, *Evaluation of the Advanced Subsonic Technology Program Noise Reduction Benefits* (Langley Research Center, Hampton, VA: NASA TM-2005-212144, May 2005), pp. 1–5.
33. Ibid., p. 4.
34. Ibid., pp. 16–17.
35. Ibid., p. 24.
36. Ibid., pp. 25–26.

This is why the smallest engine combination showed the least improvement.[37] Regardless of the actual benchmark results (which likely had greater significance to Headquarters than industry), joint industry-NASA research resumed in 2000 with the Quiet Technology Demonstrator (QTD) program. QTD tested a Boeing 777 with a Rolls-Royce Trent 800 engine incorporating nacelle and nozzle modifications. These tests, like the original AST tests, yielded significant noise reductions.[38] A second project, QTD2, partnered NASA, Boeing, General Electric, and the Goodrich Corporation. QTD2 used a GE90 (the class used in the AST research) with modified nacelles and nozzles. The landing gear included fairings to reduce airframe noise.[39] Some of these technologies were to see their operational use on the Boeing 787.[40]

High-Speed Research

HSR as a program idea dated to the 1980s, although its technical roots stretched back to the original SST program of the 1960s. It was another attempt at designing an American supersonic commercial aircraft, albeit one that overcame the environmental and economic roadblocks that led to the cancellation of earlier work. Within NASA, and especially at Langley, support for supersonic commercial aircraft had never gone away. There was always a contingent hoping to revive American efforts. A window opened in the early 1980s when an advisory committee to the White House's Office of Science and Technology Policy called on NASA to take a bolder approach to the nation's aeronautics future.[41] NASA's Aeronautics Directorate initiated a series of design studies, calling on McDonnell Douglas and Boeing to examine the feasibility of a commercially viable high-speed civil transport (HSCT). Simultaneously, Reagan's DARPA-inspired NASP came at NASA from another angle. NASP lent momentum to the civil transport, though NASA's engineers did not give much credence to a hypersonic commercial passenger vehicle. Nevertheless, all speed ranges above

37. Ibid., pp. 98–100.
38. Peter Bartlett et al., "The Joint Rolls-Royce/Boeing Quiet Technology Demonstrator Programme" (10th AIAA/CEAS Aeroacoustics Conference, Manchester, England, 10–12 May 2004 [AIAA 2004-2869]).
39. William H. Herkes, Ronald F. Olsen, and Stefan Uellenberg, "The Quiet Technology Demonstrator Program: Flight Validation of Airplane Noise-Reduction Concepts" (12th AIAA/CEAS Aeroacoustics Conference, Cambridge, MA, 8–10 May 2006 [AIAA 2006–2720]).
40. Bob Burnett, "Ssshhh, We're Flying a Plane Around Here," *Boeing Frontiers* 4, no. 8 (December 2005 January 2006), *http://www.boeing.com/news/frontiers/archive/2005/december/ts_sf07.html* (accessed 28 March 2012).
41. Conway, *High-Speed Dreams*, pp. 204–205.

(NASA image EL-1998-00237)

Figure 6.2. High Speed Research model in Langley's 14-by-22-foot tunnel, 1995.

Mach 1 were on the table. Ultimately, the contractors and Langley's own group all concluded that a supersonic aircraft traveling between Mach 2 and Mach 3.2 was viable and could enter service around the turn of the century, as long as the environmental problems could be solved. Of these, there were three: airport noise, sonic booms, and atmospheric ozone depletion.[42]

Congress and the White House green-lighted HSR for fiscal year 1990, letting NASA proceed with an initial phase that examined whether researchers could overcome the environmental roadblocks. The airport noise problem arose because the best engines for supersonic speeds were low-bypass turbofans. Without remediation, the hot, high-speed exhaust mixed turbulently with the cool outside air. General Electric and Pratt & Whitney, the two engine manufacturers, explored different ways to more smoothly mix the exhaust with the outside air. Another team looked at different engine designs, searching for one that did not have such hot, fast exhaust in the first place. In the end, they decided that a "mixed flow" turbine with a mixed nozzle would achieve the

42. Ibid., pp. 213, 216–220.

noise reductions required.[43] On the matter of sonic booms, the researchers took two approaches, hoping to find out what level of boom was acceptable to the public (none) and how much the boom could be reduced. The outcome of this was that the sonic boom problem was not solved (i.e., even after technical mitigation, people still perceived the booms as too loud). However, the economic viability of a purely over-water HSCT (where sonic booms would be heard only by marine animals and ocean liners) meant that HSCT could continue.[44] Finally, on the question of damage to the ozone, the teams used a combination of atmospheric measurements, databases and predictive models, and tests of engine combustors to determine that an ozone-neutral HSCT was possible.[45]

With a sufficient level of technical confidence achieved in phase 1, NASA began lobbying for the riskier and costlier phase 2, which sought to develop the key technologies that would make the HSCT possible. Making the phase 2 budget more palatable was an agreement from the industrial partners to help fund the research. This not only reduced NASA's costs but was also a litmus test for confidence in the project; industry was putting its money where its mouth was. Phase 2 moved forward in 1993, receiving support from industry, Goldin, and OMB.[46] The main technological sticking points tended to be in materials, which, for both the airframe and the engines, needed to be both lightweight and able to withstand very high temperatures. The program did a broad search of lightweight, high-temperature-compatible materials for the airframe and settled on a composite called PETI-5. The engines and the nozzles likewise required special attention in design and materials, especially the combustors. Thrust, weight, emissions, and operational durability all had to hit their targets, or the aircraft would not make money.[47] Although Langley, Lewis (now Glenn), and the industrial partners were making real progress chipping away at the technical issues, they were not keeping to their schedule. The timeline for entry into production was stretching out into the future. More time and more money would be needed than originally were planned for phase 2. Complicating the program was the likelihood that noise restrictions would be stricter by the time the HSCT entered service, an ominous turn for a technol-

43. Ibid., pp. 221–222, 242–247. A "mixed flow" turbofan bleeds off air after the compressor stage (thus bypassing the combustion and turbine stages), recombining it with the fast, hot core exhaust before exiting. This is distinct from high-bypass turbofans in which the bypass occurs after the fan but *before* the compressors.

44. Ibid., pp. 247–253.

45. Ibid., pp. 227–241.

46. Ibid., pp. 226–227, 253–257.

47. Ibid., pp. 264–271.

ogy that was using every means possible to meet existing regulations. Boeing, considering the noise issue and concerns about engine performance figures, lost enthusiasm for the project. Additionally, top-level support at NASA began to wane as the Space Station consumed more of the Agency's focus and budget. Not long thereafter, in 1998, HSR came to a close.[48]

Following the 1990s, there was a very modest revival of supersonic research, this time for a business jet-sized aircraft that would serve a very exclusive market segment. Of particular importance would be the sonic boom abatement measures. Beyond this, the large commercial manufacturers made no plans for supersonic transports. Boeing briefly entertained a sleek aircraft it called the "Sonic Cruiser," but this would have flown only in the high subsonic range.

UH-60 Airloads

The UH-60 Airloads program, a rotorcraft project jointly funded by NASA and the U.S. Army, was in terms of funding almost trivial compared to the vast AST and HSR programs. It turned out to be one of the two most important rotorcraft programs of the NASA era, the other being the XV-15 tilt-rotor.[49] Where the tilt-rotor attacked the developmental problems of a creating a practical vehicle, the Airloads program was a fundamental baseline study that produced a new and valuable performance dataset. The Airloads data would be at the heart of a whole new generation of rotorcraft research, especially more accurate CFD models of blade and helicopter performance. Despite the subsequent importance of the project, it lost funding even before it had begun data collection. It was a victim of both a prolonged gestation period and external budget pressures.

The original vision in the early 1980s was to conduct a battery of flight tests, wind tunnel tests, scale-model tests, and flutter tests across a variety of rotors. Called the "Modern Technology Rotors Program," its researchers planned to generate new performance data, updating work that had been done two decades earlier with older generations of helicopter blades. The best known of the earlier studies was work done by James Scheiman at Langley on an H-34 rotor system in mid-1960s. The H-34 rotor was also tested in the Ames 40-by-80-foot tunnel, providing additional data for cross-referencing. The H-34, however, was not a modern helicopter and was hardly cutting-edge when Scheiman ran his tests. The H-34 first flew in the mid-1950s and was powered by a radial piston engine. The Modern Technology Rotors Program

48. Ibid., pp. 284–285, 293.

49. Wayne Johnson, telephone interview by Robert G. Ferguson, 30 May 2007, NASA HRC; William G. Warmbrodt, telephone interview by Robert G. Ferguson, 14 May 2007, NASA HRC.

(NASA image AC93-0010-34)

Figure 6.3. Sikorsky UH-60 Airloads helicopter flying above Ames Laboratory, 1993.

would not only use rotorcraft with much newer blades but would benefit from advanced instrumentation. The Sikorsky UH-60A, the new workhorse for the military, became the program's first test subject.[50]

50. William G. Bousman, "UH-60A Airloads Program," Occasional Note 1999–2001, 29 March 1999, Ames Research Center (ARC); Michael E. Watts and Jeffrey L. Cross, "The NASA Modern Technology Rotors Program" (AIAA/AHS/CASI/DGLR/IES/ISA/HEA/SETP/SFTE 3rd Flight Testing Conference, Las Vegas, NV, April 1986); Johnson interview, 30 May 2007; Chee Tung, telephone interview by Robert G. Ferguson, 20 April 2007, HRC; W. J. Snyder, "Rotorcraft Flight Research with Emphasis on Rotor Systems," in *NASA/Army Rotorcraft Technology*, vol. 3, *Systems Integration, Research Aircraft, and Industry* (proceedings of a conference sponsored by the

In 1985, NASA and the Army contracted with Sikorsky for a special set of UH-60A blades. One blade would have 242 pressure transducers to measure the air pressure around the blade's airfoil, and the other would have strain gauges and accelerometers. It was 1988 before Sikorsky delivered the blades. Unfortunately, it took the researchers and technicians at Ames another four years before they had a working helicopter. The problem was integrating all the spinning instrumentation and reliably collecting the data as they came in from the sensors. By the close of 1992, the Ames team finally had a working Rotating Data Acquisition System, so flight testing was set for the following summer. In May 1993, the program lost its funding, but with the fiscal year ending 30 September, there was still time to squeeze in some flight testing covering four program flight objectives. Scrounging for funds elsewhere, NASA was able to support continued flights through February 1994, covering an additional 26 test objectives. In total, the program yielded some 36 gigabytes of test data (a vast quantity at the time).[51]

The Airloads program was a far cry from what had been envisioned originally. Instead of seven or more rotorcraft, it ended up with one. Instead of a battery of tests (flight tests, wind tunnel, etc.), the program ended up with only flight-test data, but the data were, according to the researchers, pure gold. Combined with maturing CFD capabilities, the data took on even greater importance, for they could be used to refine and validate predictive codes.[52] Funding for the program came entirely out of the base rotorcraft budget and, over the life of the program, was estimated to be on the order of $10 million.[53]

In spite of the eventual success of the Airloads program, rotorcraft research took a grave turn in the following decade. NASA decided to discontinue support for all rotorcraft activities for fiscal year 2002, and the Agency closed the critical 40-by-80-by-120-foot wind tunnel in 2003. Aeronautics was certainly under immense budgetary pressure, but rotorcraft lacked sufficient

Department of the Army and NASA, Ames Research Center, Moffett Field, CA, 17–19 March 1987, NASA CP-2495), pp. 1234–1273.

51. Although 36 gigabytes is not large by contemporary standards, it was a large dataset at the time and required special hardware for storage. Bousman, "UH-60A"; Warmbrodt interview.

52. In the words of rotorcraft researcher Wayne Johnson, "UH-60 Airloads program…is now considered a premier database and is driving the CFD development throughout the U.S." See Wayne Johnson interview. The Airloads program data are described in William G. Bousman, "UH-60A Airloads Program," Occasional Note 1999–2002, 3 May 1999, ARC.

53. Warmbrodt interview.

political support at Headquarters.[54] NASA also might have been hoping that the Department of Defense would take up the slack. By 2004, a National Research Council committee, invited by NASA to evaluate the Agency's aeronautics program, made the reconstitution of rotorcraft research 1 of 12 top-level recommendations.[55] In 2006, NASA did just that, and to the good fortune of the rotorcraft programs, the U.S. Air Force took over the operation of the 40-by-80-by-120-foot wind tunnel with a 25-year lease beginning in 2006.[56] The following year, NASA and the Department of the Army penned a new memorandum of understanding regarding aeronautics, thus reconfirming NASA's partnership in rotorcraft research.[57]

General Aviation, Small Turbines, and SATS

NASA has had a long association with the general aviation (GA) segment of the commercial aircraft market. General aviation refers to nonmilitary, non-scheduled aircraft flights. It comprises both commercial and noncommercial activities and is, for many of the nation's smaller community airports, their primary activity. Although an important economic activity, general aviation has been of tertiary importance behind national defense and scheduled commercial travel. Because larger aircraft are already studied under NASA's military and subsonic research programs, GA-specific research has focused on smaller aircraft, i.e., business jets to personal aircraft.

The GA market encompasses a number of loosely related segments. The light aircraft category, which grew significantly in the post–World War II decades, includes small single- and twin-engine models such as Cessna, Piper, and Beech aircraft. A dedicated business aircraft market grew in the late 1950s and early 1960s with Grumman's Gulfstream division, North American's Sabreliner, and Learjet leading the way. There are niche markets as well, such as bush

---

54. See NASA Aeronautics annual budgets, 2002 to 2004, in historical materials for *NASA's First A*, NASA HRC.
55. National Research Council, Committee for the Review of NASA's Revolutionize Aviation Program, "Review of NASA's Aerospace Technology Enterprise: An Assessment of NASA's Aeronautics Technology Programs" (National Academies Press: Washington, DC, 2004), p. 8.
56. Tim White and Andy Roake, Arnold Engineering Development Center Public Affairs, U.S. Air Force Materiel Command News Service, "National Full-Scale Aerodynamics Complex Set To Reopen," press release, 9 April 2006.
57. "Memorandum of Understanding between the Department of the Army and the National Aeronautics and Space Administration concerning Collaborative Research in Aeronautics," executed in July and August of 2007 by Michael D. Griffin, NASA Administrator, and Pete Green, Secretary of the Army.

(NASA image L-74-2499)

Figure 6.4. A general aviation airframe hangs, awaiting a crash landing, in Langley's massive Impact Dynamics Research Facility in 1979.

pilots and crop-dusters. Finally, a home-built market has advanced since the 1970s, relying heavily on kit-built designs.[58] Reflecting this diverse makeup, the main political voices of GA have included the small aircraft manufacturers, the National Business Aviation Association, the Airline Owners and Pilots Association, and the Experimental Aircraft Association.

Much of NASA's research associated with general aviation has typically been applied rather than fundamental. A good example is Langley's GA crash testing performed over a 30-year period beginning in the 1970s. Such testing examined the survivability of existing aircraft, the verification of predictive computer models, the performance of new materials, and new design techniques. At the center of these tests was the Impact Dynamics Research Facility (IDRF), a 240-foot-high gantry built in the 1960s as part of a lunar lander simulator. After the Apollo program, Langley reconfigured the gantry to allow for controlled crashes of small fixed-wing aircraft and helicopters. The initial program,

58. Donald M. Pattillo, *A History in the Making: 80 Turbulent Years in the American General Aviation Industry* (New York: McGraw-Hill, 1998).

the General Aviation Aircraft Crash Test Program, was a cooperative effort that brought together NASA, the FAA (as the regulatory body for aircraft certification and safety procedures), and manufacturers. Starting in 1974 and lasting a decade, the program crashed 33 full-scale aircraft. As with the wind tunnels, researchers compared actual crash results with computational simulations, thus refining mathematical means for predicting structural performance. The facility saw occasional use after this initial program, with an additional eight full-scale crash tests completed by the time of the IDRF's closure in 2003.[59]

Another post-Apollo effort was Lewis's investigations of general aviation propulsion, notably small turbines. Lewis focused on space propulsion in the 1960s, so this was a return to atmospheric propulsion in the same vein as the advanced turboprop work discussed in chapter 5. Lewis oversaw the General Aviation Turbine Engine (GATE) program and the Quiet, Clean General Aviation Turbofan (QCGAT) program. The GATE program sought to address the dearth of small turbines (and small turbine research at NASA) in the GA size range, that is, 1,000 shaft horsepower and below. The program underwrote independent paper studies at four companies: Garrett AiResearch, Detroit Diesel Allison, Teledyne CAE, and Williams Research. The participants paid special attention to ways in which small turbines could be produced more economically. All of the participants chose turboprops as the preferred configuration, although each employed different means for reducing costs. Williams, for example, designed the engine (especially the turbine blades) to experience low loads and predicted that this would allow them to use lower-cost components and manufacturing processes. QCGAT sought to apply the lessons learned from NASA large jet turbine programs (i.e., QCSEE) to small turbines. AiResearch and Avco Lycoming both delivered test engines to Lewis in 1979. The QCGAT program used existing engine cores, thus reducing cost but also reducing the potential for dramatically smaller engines. The two engines met all of the noise goals, most of the emissions goals, and some of the performance goals.[60]

In the early 1980s, the market for light aircraft (i.e., primarily piston-engine models) crashed. Sales peaked in 1978 and 1979 (just as GATE and QCGAT were under way) with over 17,000 shipments each year and then dropped precipitously. By 1983, GA shipments were below 3,000; the industry reached

59. Karen E. Jackson et al., "A History of Full-Scale Aircraft and Rotorcraft Crash Testing and Simulation at NASA Langley Research Center" (4th Triennial International Aircraft and Cabin Safety Research Conference, Lisbon, Portugal, 15–18 November 2004).

60. *General Aviation Propulsion* (conference proceedings, Lewis Research Center, Cleveland, OH, 28–29 November 1979, NASA CP-2126).

bottom in 1992 at 899 shipments. The impact on the light aircraft industry was predictably dramatic. Cessna ceased production of piston-engine aircraft in 1986. Beech closed factories. Piper declared bankruptcy in 1991. Numerous other manufacturers, such as Bellanca, Fairchild, and Rockwell, fell by the wayside or departed the field. Cessna and Beech survived, in part, because they were no longer independent companies; Raytheon purchased Beech in 1980, and General Dynamics purchased Cessna in 1985 (and subsequently sold it to another conglomerate, Textron, in 1992). Both manufacturers shifted their product lines toward the growing business jet market.[61]

There were a number of causes for the market collapse. Some observers pointed to a lack of innovation by traditional GA manufacturers or to the indirect effects of airline deregulation. The manufacturers pointed their fingers squarely at the rising cost of aviation litigation. Not only were aviation lawsuits on the rise through the 1970s, but general aviation aircraft also had long lifespans. Manufacturers were finding themselves the target of lawsuits for aircraft produced decades earlier, and, with each passing year, the burden only increased. Liability insurance had become a significant fraction of each new aircraft sold, sometimes doubling the cost. Furthermore, these new aircraft had to compete against a vibrant used-aircraft market.

As traditional manufacturers receded from the market, home-built designs surged. The home-built market benefited not simply from its do-it-yourself character, but also from FAA rules that reduced upfront and operating costs as long as owners built at least 51 percent of the aircraft. Aircraft sold as kits and assembled by owner-builders did not require that manufacturers carry liability insurance. Further, the FAA allowed home-builders to perform their own maintenance. On top of the cost advantages, the home-built sector was strongly experimental, with designers such as Burt Rutan pushing novel materials (composites) and configurations (canards) that gave vitality to the home-built segment. By comparison, the traditional light aircraft manufacturers focused on incremental improvements to tried-and-true designs and fabrication methods.

It was not until the 1990s that the government moved to address the problems of the GA industry. Congress passed, and President Clinton signed, the General Aviation Revitalization Act (GARA) in 1994, which protected manufacturers from liability for aircraft older than 18 years. After GARA, Cessna reentered the piston-aircraft market. The delivery of piston aircraft accelerated,

61. Pattillo, *A History in the Making*, pp. 125, 152; U.S. General Accounting Office, "General Aviation: Status of the Industry, Related Infrastructure, and Safety Issues" (Report to Congressional Requesters, August 2001).

but it did not approach the levels reached two decades prior.[62] The FAA also worked to stimulate the GA market, planning in the 1990s for a new category of aircraft and pilot licensing called, respectively, the Light Sport Aircraft category and the Sport Pilot Certificate. Released in 2004, these programs attempted to simplify regulations and encourage the manufacture of a new class of cost-sensitive aircraft and the growth of a new generation of pilots.[63]

NASA, for its part, reorganized and reenergized its GA research. The centerpiece of the 1990s was the Advanced General Aviation Transport Experiments (AGATE) program. AGATE sought to provide technologies meant to accompany the GA renaissance. Under AGATE, Langley expanded the crash-test program to include new aircraft, some of them composite designs, some of them from newer manufacturers that had begun as kit-aircraft businesses.[64]

Lewis, as part of its role in AGATE, reprised its small turbine support with a General Aviation Propulsion (GAP) program. Through the 1980s, Williams Research, now Williams International, continued work on small turbines and found a successful design in its FJ44 engine. Cessna used the Williams engine in its expanding line of economical business jets. Under GAP, Lewis sponsored a four-year program at Williams to produce an even smaller turbine, one appropriate for a four- to six-person aircraft. The result was the FJX-2. Williams attempted to nurture interest in the engine by demonstrating smaller turbines in prototype jets, dubbed V-Jets. The V-Jet II, in fact, was an all-composite aircraft designed and built by Burt Rutan's Scaled Composites.[65]

From all of this sprang a new category of aircraft, the very light jet, or VLJ. In 1998, a new company called Eclipse Aviation opened. The company's business plan was to take the V-Jet II and, through lean manufacturing techniques, produce the design in high numbers and at low cost. The company hoped to

---

62. GAO, "General Aviation: Status of the Industry, Related Infrastructure, and Safety Issues"; Adam Bryant, "Aviation Bill Encourages Manufacturers," *New York Times* (4 September 1994); Adam Bryant, "Spending It; Small Planes Are Coming Back," *New York Times* (19 March 1995): Business section.
63. Department of Transportation, Federal Aviation Administration, "14 CFR Parts 1, 21, et al., Certification of Aircraft and Airmen for the Operation of Light-Sport Aircraft; Final Rule," Federal Register 69, no. 143 (27 July 2004): 44772–44882.
64. Jackson, "A History of Full-Scale Aircraft and Rotorcraft Crash Testing and Simulation."
65. "General Aviation Propulsion (GAP) Program, Turbine Engine System Element" (Performance Report, NCC3-514, 6 October 1997); Williams International, "The General Aviation Propulsion (GAP) Program" (Walled Lake, MI: NASA CR 2008-215266, July 2008). This latter document appears to have been authored in March 2002 and covers almost exclusively Williams International's development of the FJX-2.

create a new group of small aircraft owners who flew their own VLJs, and they also sought to spur the rise of VLJ air taxi services.[66] A number of competing manufacturers entered, or attempted to enter, the same market space over the next decade, including Adam Aircraft Industries with its A700, Cessna with its Model 510 Mustang, and Piper with its Piperjet.[67] Williams's FJ33 engine found competition from Pratt & Whitney, which brought its small turbofan to the market early the following century. The VLJ market, unfortunately, existed in large part on credit that was financing both the new manufacturers and the air taxi operations. When the market for VLJs did not grow as fast as anticipated, and when the credit markets seized in 2008, only established operations remained standing. Both Eclipse and Adam Aircraft entered bankruptcy, as did some VLJ air taxi operations.[68]

The VLJ market, if not as large as hoped, became an established category in no small measure because of the developmental engine efforts of NASA. NASA partnered well and nurtured the technology in step with the manufacturers, but NASA researchers also began to cast about for ways in which to make flying more accessible. General aviation's popularity and usage was limited in part by the cost of owning or renting an aircraft, as well as the high training requirements and safety concerns. As the AGATE program drew to a close in 2001, Langley's general aviation researchers began a successor program that focused on creating system-level technologies to make flying small aircraft easier and safer. This was the Small Aircraft Transportation System, or SATS.[69]

As a program, SATS adopted what had worked well under AGATE, namely, the close partnering with industry, universities, and trade associations. Partnering had numerous advantages. Technically, it eased the two-way flow of information between researchers in industry and NASA. This was especially important for a developmental project that sought to stimulate innovation among the manufacturers. Further, there were political advantages since bringing many companies on board across a wide geographic area naturally increased (but did not assure) the program's political backing. With SATS, Langley and

66. Joe Sharkey, "On the Road; Standing on a Runway, Hailing an Air Taxi," *New York Times* (28 February 2006): Business section.

67. This is not an exhaustive list. Through the early 21st century, a number of companies fielded and continued to work on VLJ prototypes. Only the Eclipse, Cessna, and a model from Embraer have made it to market.

68. Joe Sharkey, "Air Taxis Fly into Financial Turbulence," *New York Times* (20 May 2008): Business section.

69. Bruce J. Holmes, "From the Desk of the Program Manager," *AGATE Flier* 6, no. 1 (July 2000): 2, 10–11. This publication came from the AGATE Alliance Foundation based in Hampton, VA.

the National Consortium for Aviation Mobility established six laboratories in the central and eastern United States.[70]

At a general level, the intention of SATS was to help create a system of "personalized air travel," one that took advantage of the country's thousands of general aviation airports in suburban and rural areas. If the AGATE program had, among other things, helped create low-cost jets that seated as many people as a family car, the SATS program sought to reduce the barriers that discouraged wide use of such aircraft. Congress and the White House had done their part in reducing manufacturer's liability. NASA and manufacturers had come out with advanced new aircraft that were safer and, arguably, more affordable than earlier models. SATS took on the next challenge: making flying as casual and easy as hopping in a car. SATS ran from 2001 to 2006, and it focused on four areas. First, SATS searched for ways for people to use the many GA airports that lacked traditional navigational aids (e.g., instrument landing systems or ILS, radar, and control towers) in conditions of low visibility. Researchers sought a standard to allow operations down to a 200-foot ceiling and half a mile of visibility. Second, they sought to increase the number of aircraft that could simultaneously use such airports. During periods when instrument flight rules were in effect, such airports were restricted to one aircraft movement (e.g., landing) at a time, greatly decreasing throughput. Third, SATS examined ways to make flying easier and safer for less proficient pilots. Finally, SATS researchers sought to make this new system integrate seamlessly within the existing airspace system.[71]

Fortunately for SATS, a new navigational technology was coming online that gave aircraft more accurate positioning information, something that would be critical for landing at airports without ILS equipment. In 2003, the FAA began operation of the GPS Wide Area Augmentation System (WAAS), which measures variations in GPS transmissions and sends deviation correction messages to GPS-WAAS–enabled receivers. For SATS, researchers took these navigational data and fed them into a terrain database that could be presented to the pilot as part of a synthetic vision system. The synthetic system could also portray airborne traffic. On top of this, the researchers installed a low-light, infrared camera that would provide an actual image of the landing strip as it came into view. Synthetic and actual imagery were thus overlaid.[72]

---

70. Guy T. Kemmerly, "The Small Aircraft Transportation System Project: An Update," *Journal of Air Traffic Control* 48, no. 2 (April 2006): 13–18.

71. Ibid.

72. Ibid.

In order to make simultaneous use of the non-instrumented airport during ILS conditions, the SATS researchers again turned to newly available technology that greatly reduced the cost of implementation. Automatic dependent surveillance-broadcast (ADS-B), which had been in the works for many years and which the FAA would begin to deploy in 2009, was a system in which aircraft continuously broadcast their state vector (position and velocity) as a data message. The SATS solution was to install a small computer at the airport (known as a Self-Controlled Area) that would accept a request to land and use ADS-B data to sequence traffic automatically. Data links between the computer and the small aircraft would send the appropriate navigational directions.[73] As for easing pilot workload and increasing safety, one of the key technologies on which researchers worked was a "highway in the sky." Whereas pilots traditionally take their navigational cues from a number of instruments, the SATS system integrated the information and presented it as a series of graphics (e.g., rectangles) on a head-up display.[74]

Taken as a whole, AGATE and SATS represent a unique attempt to strongly influence a particular market segment. True, the SST (and subsequent high-speed efforts) as well as the STOL programs sought to create new classes of commercial operation, but AGATE and SATS differed in that they succeeded in nurturing a new class of aircraft, spurred by long-running support for small turbines (dating to the 1970s). Even though NASA's STOL programs proved the validity of various technologies, the SATS program went further than similar STOL efforts in the design and operation of new air traffic control arrangements, something that was once considered largely an FAA domain. This was, certainly, a reflection of NASA's more recent push into air traffic control research generally. When combined with the FAA's regulatory changes and Congress's limits on manufacturer's liability, this was a remarkable, multifaceted government effort to support an industry.

Conclusion

The grandiose nature of AST and HSR, along with their collapse, makes it difficult to view the decade as anything but deeply troubled. It was as if aeronautics

73. Maria C. Consiglio, Victor A. Carreño, Daniel M. Williams, and César Muñoz, "Conflict Prevention and Separation Assurance Method in the Small Aircraft Transportation System" (AIAA 5th Aviation Technology, Integration, and Operations Conference, Arlington, VA, 26–28 September 2005, NASA LF99-1688).
74. Kemmerly, "The Small Aircraft Transportation System Project," pp. 13–18; "Small Aircraft Transportation System Aircraft Technologies Lead Way to Future Flight," *AGATE Flier* 5, no. 1 (March 1999): 4.

finally got what it wanted: good funding increases, strong political support, and outsize programs to match the rest of NASA. This level of support proved shallow, and when push came to shove, aeronautics fell spectacularly. The cost to the Agency was not merely the demise of two oversize programs, but the loss of continuity in healthy research communities. Finally, the rotorcraft program's loss of funding for the Airloads research (and the later elimination of rotorcraft funding altogether) illustrated the kind of capricious budget tragedy that could strike productive programs.

The decade's funding woes should not color our evaluation of the technical accomplishments. The carbon-stitched wing, in spite of not being adopted by the aircraft manufacturers, did work. While we may view cynically the close relationship and subsidized capital investment at McDonnell Douglas, NASA was operating exactly as its leadership intended: in close coordination with, and at the service of, its customers. The quiet engine research returned useful techniques that saw their way into the next generation of jet engines and prompted follow-on research in the next decade. Finally, the UH60 Airloads program showed that there was still critical fundamental research to be done.

Chapter 7:

Caught in Irons

If there were ever a point at which the future of NASA's aeronautics research appeared in doubt, the turn of the millennium was it. Aeronautics had a difficult time of it in the 1990s: drastic funding cuts imperiled the health of research programs while NASA's raison d'être was undercut by the end of the Cold War. At least NASA had, in the Clinton White House, a proponent of federally funded research. The incoming President, however, had no such inclination. President George W. Bush's administration held that in the absence of market discipline, government was vulnerable to waste, fraud, and abuse. Federally supported R&D, specifically, was found to suffer from vague goals, insufficient performance monitoring, uncontrolled program growth, and unnecessary competition with and/or subsidization of private R&D.[1] Greatly complicating an already tight Agency budget, President Bush called in January 2004 for the country to return to the Moon by 2020 as a stepping-stone to sending humans to Mars.[2] The buildup to the Mars mission was to be accomplished in parallel with a resumption of Shuttle flights (grounded after the February 2003 loss of Space Shuttle Columbia) and the completion of the International Space Station, all within a budget that increased only slightly faster than inflation.

Within the Agency, leadership changed hands in 2001 when Sean O'Keefe replaced Dan Goldin as Administrator, and again in 2005, when Dr. Michael Griffin replaced O'Keefe. A seemingly constant top-level restructuring of the focus of the aeronautics mission followed the termination of the High Speed Research and Advanced Subsonic Technology programs. From the late 1990s to 2005, Headquarters promulgated new Agency goals for itself and aeronautics. These goals were well meant, but with each shift and funding realignment came the impression that the Agency's aeronautics mission had become unmoored. What was a scientist or engineer to make of goalposts that moved every two or

1. Office of Management and Budget, Executive Office of the President, "The President's Management Agenda," 2002, pp. 2–4, 43.
2. David E. Sanger and Richard W. Stevenson, "Bush Backs Goal of Flight to Moon To Establish Base," *New York Times* (15 January 2004): section A, p. 1.

three years, especially when a research project could take substantially longer to design, fund, and execute?

To understand these program shifts, it is worthwhile to review Dan Goldin's tenure. In a March 1997 speech to the Aero Club of Washington, Goldin gave the Aeronautics and Space Transportation Enterprise his Three Pillars of Success and 10 goals. The Pillars were Global Civil Aviation, Revolutionary Technology Leaps, and Access to Space. The 10 goals, 8 of which related to aeronautics, were reportedly given to him by the aeronautics enterprise. Certainly meant to inspire, the goals were so optimistic that, in hindsight, they strike an outlandish tone. For example: the national airspace system was to triple its capacity in 10 years; aircraft accidents were to be reduced by a factor of five within 10 years; the cost of air travel was to be reduced by 25 percent within 10 years and 50 percent in 25 years; and travel time on transpacific and transatlantic flights was to be reduced by half within 25 years with no cost increase.[3] At about the same time, the Aeronautics Technology Enterprise merged with the Space Technology Enterprise to form the Aeronautics and Space Transportation Technology Enterprise. The new structure took on Goldin's Three Pillars as its mission statement and stressed that "Although we do not know in advance how to achieve the goals and objectives, the development of investment strategies is issue driven."[4] Aeronautics, thus, was being organized top-down around a set of aggressive technical goals.

For 1999, the Enterprise changed names again to become the Aero-Space Technology Enterprise (simplified to Aerospace the following year). In 2000, the Three Pillars were replaced with 4 goals and 11 objectives. The goals were to revolutionize aviation, advance space transportation, pioneer technology innovation, and commercialize technology. These goals persisted into the 2001 Aerospace Technology Enterprise strategic plan.[5] The next year, with the Agency now under Sean O'Keefe and operating in a post-9/11 environment,

---

3. NASA, *Spinoff 1997* (Washington, DC: NASA NP-1997-08-226-HQ), p. 12; "Aeronautics and Space Transportation Technology: Three Pillars for Success," brochure, 1997; Robin McMacken, "Whitehead Unveils Pillars," *Dryden X-Press* (16 May 1997); Daniel S. Goldin, "The Three Pillars of Success for Aviation and Space Transportation in the 21st Century" (speech to the Aero Club of Washington, American Institute of Aeronautics and Astronautics, and National Aviation Club, Washington, DC, 20 March 1997).
4. NASA Strategic Plan 1998 (Washington, DC: NASA Policy Directive [NPD] 1000.1), p. 31.
5. NASA Strategic Plan 1998 with 1999 Interim Adjustments (Washington, DC: NASA NPD 1000.1a); NASA Strategic Plan 2000 (Washington, DC: NASA NPD 1000.1b, September 2000); Aerospace Technology Enterprise Strategic Plan, April 2001; NASA Aerospace Technology Enterprise, Annual Progress Report, "Turning Goals into Reality" (NP-2001-04-265-HQ, 2000).

aeronautics received its own blueprint with four focus points: digital airspace, revolutionary vehicles, security and safety, and state-of-the-art educated workforce. Aeronautics was split further into three areas: aviation safety and security, airspace systems, and vehicle systems. Also in 2002, NASA promulgated three broad themes with 10 goals. The first theme echoed the rhetoric of national security concerns, but on a larger scale: "Understand and Protect Our Home Planet." Of the 10 goals, 3 applied to aeronautics:

- Enable a safer, more secure, efficient, and environmentally friendly air transportation system
- Create a more secure world and improve the quality of life by investing in technologies and collaborating with other agencies, industry, and academia
- Enable revolutionary new capabilities through new technology

Lest anyone find him- or herself confused by the shifting goals, the 2002 Annual Progress Report for the Enterprise provided a chart that mapped the old 2002 Enterprise goals and objectives with the 2003 Agency goals and objectives. The author's note says, "In most cases, there is a direct correlation between the 2002 and 2003 goals and objectives, though in some cases assignments have changed and the new relationship is less clear."[6] By the time the 2003 Aerospace Technology Enterprise Strategy was published, aeronautics research areas had been carefully mapped onto the Agency's goal/objective matrix. While it was apparent that aeronautics was supporting aspects of the Agency's most recent vision, it is not at all apparent that the vision was guiding aeronautics R&D decisions. Indeed, there appeared to be a distinct disconnect.[7]

The constant institutional realignments had a destabilizing effect that did not go unnoticed outside the Agency. In 2003, the National Research Council (NRC) published an evaluation of aeronautics' Pioneering Revolutionary Technology (PRT) program done by the NRC's Aeronautics and Space Engineering Board. The NRC noted:

> Changes in priority, organization, and funding will always occur and should be expected in a dynamic research program. However, the PRT program has undergone frequent and sometimes disruptive

---

See also NASA Aerospace Technology Enterprise, Annual Progress Report, "Turning Goals into Reality" (NP-2002-04-287-HQ, 2001).

6. NASA, "Aeronautics Blueprint—Toward a Bold New Era of Aviation" (Washington, DC: NASA NP-2002-04-283-HQ, 2002); NASA Aerospace Technology Enterprise, Annual Progress Report 2002 (Washington, DC: NASA NP-2003-06-306-HQ, 2003).
7. NASA Aerospace Technology Enterprise Strategy (Washington, DC: NASA NP-2003-01-298-HQ).

> restructuring and reorganization. Some of these changes appeared to be a destructive force rather than a natural reallocation of resources as part of research progress and maturation.... The committee recognizes that certain time spans are imposed by the Office of Management and Budget. However, the OMB constraints apply 5-year time horizons, whereas the past incarnations of the PRT program experienced reorganization at 1- and 2-year intervals. Even during the course of this 12-month review, portions of the PRT program were renamed and other portions reorganized in significant ways.[8]

With the appointment of Michael Griffin as NASA Administrator in 2005, aeronautics underwent a thorough reorganization. One of Griffin's major changes was to greatly curtail the use of visionary goals as organizational planning tools. NASA's strategic goals were simplified and made concrete (e.g., "Strategic Goal 1: Fly the Shuttle as safely as possible until its retirement, not later than 2010"). There was no mapping exercise tracing prior goals to new goals. Aeronautics was given its own organizational standing as the Aeronautics Research Mission Directorate (ARMD), one of four directorates within the Agency. The plan noted, "In recent years, the emphasis on transferring technologies to end-users shifted NASA's focus from long-term, high-risk, cutting-edge research to short-term technologies and 'point solutions' to complex challenges."[9] ARMD was given four objectives:

- Reestablish NASA's commitment to mastery of core aeronautics competencies in subsonic (rotary and fixed-wing), supersonic, and hypersonic flight
- Preserve the Agency's research facilities as national assets
- Focus research in areas that are appropriate to NASA's unique capabilities
- Directly address the needs of the Next Generation Air Transportation System

While aeronautics retained its Airspace Systems program and the Aviation Safety and Security program became simply Aviation Safety, Vehicle Systems was refashioned into the Fundamental Aeronautics program. A new program,

8. National Research Council, Committee for the Review of NASA's Pioneering Revolutionary Technology (PRT) Program, Aeronautics and Space Engineering Board, "Review of NASA's Aerospace Technology Enterprise: An Assessment of NASA's Pioneering Revolutionary Technology Program" (Washington, DC: National Academies Press, 2003).
9. NASA Strategic Plan 2006 (Washington, DC: NASA NP-2006-02-423-HQ, 2006).

Aeronautics Test, was established with the sole purpose of supporting the Directorate's research infrastructure.[10]

Griffin's Agency reorganization, which ostensibly was done following President Bush's 2004 "Vision for U.S. Space Exploration," sought to reverse a number of trends that had begun under Goldin.[11] Instead of a businesslike, customer- and issue-oriented enterprise, aeronautics at its heart was to support fundamental research in four regimes (subsonic fixed, subsonic rotary, supersonic, and hypersonic). This was a 180-degree shift since it emphasized the need to support particular fields of research, regardless of the problem at hand. The organization also thinned the ranks of management and, along with giving aeronautics its own directorate, elevated the program within the Agency. One trend that did not substantially change was the decreasing budgetary commitment to aeronautics.

The inherent advantage of Griffin's approach was that research programs did not necessarily need to be realigned every time political winds shifted, nor was an institutional commitment to fundamental research susceptible to the same risks that plagued the high-profile technology programs of the 1990s. Certainly, fundamental research programs could undertake high-risk endeavors, but that was distinct from an organizational structure that tied itself to revolutionary goals. The flip side of organizing around fundamental research, however, was the lack of programmatic appeal, especially when measured against the exploits of the space program. The lack of an Apollo-like mission to rally the Agency was exactly the charge leveled decades earlier. Fortunately, Griffin's reorganization of aeronautics was assisted by an executive order from President Bush establishing a National Aeronautics Research and Development Policy in December 2006. (The reorganization had taken place at the beginning of the year.) Ever since President Carter, there had been an explicit space policy with periodic updates from subsequent administrations.[12] The aeronautics R&D policy was a step toward placing aeronautics on a similar footing, just as Griffin had given aeronautics its own mission directorate. The executive order stated that the OSTP's National Science and Technology Council was to write the policy document in order to guide R&D through the year 2020. With Griffin's administration strongly represented on the policy committee, it was no surprise that the Council's ultimate document buttressed

10. Ibid.

11. President George W. Bush, "A Renewed Spirit of Discovery: The President's Vision for U.S. Space Exploration," January 2004; NASA Strategic Plan 2006.

12. Presidential Directive/NSC-37, "National Space Policy," 11 May 1978.

the organizational changes that Griffin and ARMD Associate Administrator Lisa Porter had already enacted.[13]

Earmarks

One of the more pernicious developments in the long-term history of NASA's budget (and the federal budget in general) was the growth of congressional earmarks. Earmarks gave members of Congress the ability to allocate portions of the budget for specific projects and contractors. The more generous view of earmarks was that this was the outcome of a democratic process; a more skeptical view took earmarks as a form of pork, a political dividend for members of Congress and their constituents.[14] Though earmarks tended to be small relative to both the aeronautics and Agency-wide budgets, earmarks effectively reduced the Agency's budget, as most earmarks were unfunded, a zero-sum gain. In an organization with a history of bootstrapping innovative ideas through frugal grassroots support, even modest earmarks could have a profoundly negative impact.[15]

Earmarks were a rarity at NASA until the mid-1990s. NASA's Inspector General noted that in the 1997 fiscal budget, there were only six earmarks, but by 2006, the number would be 199. In the same period, the budget outlay for earmarks would grow from under $100 million per year to $568 million. Looking specifically at the aeronautics budget, of the $884 million provided in FY 2006 (under full-cost accounting), there were 17 earmarks totaling $97.5 million, or more than 11 percent of the Aeronautics budget.[16] What earmarks represented, more than anything else, was a new and remarkable degree of politicization of research funding. Specific research choices were now being made at a level higher than even the NASA Headquarters management. While some earmarks paralleled existing work at the Centers, some did not. Regardless, earmarks took Centers out of the loop, leaving them in some cases with unsolicited (and not competitively bid) research proposals. Center

13. Executive Office of the President, National Science and Technology Council, "National Aeronautics Research and Development Policy," December 2006; Report to Congress, "NASA Response to the National Aeronautics R&D Policy," February 2007.
14. From 1995 to 2007, earmarks for the entire federal budget increased threefold in volume to $64 billion. Jacqueline Palank, "Top Democrat Plans Advance List of Earmarks," *New York Times* (12 June 2007): section A, p. 19.
15. To see the impact of earmarks from a research center perspective, see the interview with Thomas A. Edwards by Robert Ferguson, Ames Research Center, 14 January 2005, NASA HRC.
16. NASA Office of Inspector General, "Audit of NASA's Management and Funding of Fiscal Year 2006 Congressional Earmarks" (Report No. IG-07-028, 9 August 2007).

attempts to challenge earmarks by actually vetting the content and quality of the proposals proved unsuccessful.[17]

An instructive example of the kind of work that earmarks supported was the DP-2 V/STOL aircraft designed by the duPont Aerospace Company (no relation to the Du Pont chemical company). Since the early 1970s, Anthony duPont, the president of the company, had been interested in developing a fixed-wing transport aircraft that could take off vertically by redirecting the thrust of two nose-mounted turbojet engines. His DP-2 was designed to carry nearly 50 troops. DuPont approached the U.S. Navy in 1986 and DARPA in 1990; both organizations found significant problems with the design and the enterprise. He was already well known in aeronautics circles as one of the advocates of the single-stage-to-orbit concept that had inspired the National Aero-Space Plane. Congressional earmarks routed funding to duPont Aerospace over the objections of DARPA. From 1988 to 2003, earmarks provided some $63 million to the project, some $7 million of which came from NASA in 2002–03.[18]

Part of what made the DP-2 earmark remarkable was that Congress appeared to bypass an existing community of V/STOL researchers. There was little attempt to genuinely incorporate NASA's R&D capabilities into the project.[19] From 2003 to 2006, duPont Aerospace tested a smaller-scale version of the DP-2 called the DP-1. In size and configuration, the DP-1 bore a resemblance to the first jet VTOL built in the United States, the Bell X-14. The X-14 first flew in 1953 and was operated at Ames until 1981. Both the X-14 and DP-1 buried two jet engines in the nose and redirected the thrust downward for VTOL operations.[20] Congress eventually ended funding for the DP-1/DP-2

17. Edwards interview, 14 January 2005.

18. Seven million dollars is not much in the context of the overall federal budget, but recall that the entire budget for the groundbreaking UH-60 Airloads program was approximately $10 million. U.S. House of Representatives, Committee on Science and Technology, Subcommittee on Investigations and Oversight, "Hearing Charter: The duPont Aerospace DP-2 Aircraft," 12 June 2007. NASA's contributions were $3 million in 2002 and $4.075 million in 2003. On duPont and the NASP, see Butrica, *Single Stage to Orbit*, pp. 78–79.

19. For the period during which NASA funds were earmarked to duPont Aerospace, NASA was charged with determining the airworthiness of the test vehicle. U.S. House of Representatives, Committee on Science and Technology, Subcommittee on Investigations and Oversight, "Hearing Charter: The duPont Aerospace DP-2 Aircraft," 12 June 2007.

20. It is also worth noting that a number of major aircraft manufacturers had undertaken preliminary design studies of midsize jet VTOL aircraft in the 1980s, some of which were the subject of joint NASA-industry analysis. See, for example, Megan A. Eskey and Samuel B. Wilson III, "The Handling Qualities and Flight Characteristics of the Grumman Design 698 Simulated Twin-Engine

in 2007, when political control shifted. In March of 2010, the House voted to ban earmarks to corporations.[21] Only time would tell whether this was a historical high-water mark for the politicization of research decision-making or merely a pause in a long-term trend.

Blended Wing Body

In 2007 at Dryden, NASA and Boeing researchers saw the fruits of more than a decade of research and campaigning go aloft in the form of the X-48B remotely piloted vehicle. The X-48B was a blended wing body (BWB) aircraft, a design hatched in the 1980s in the private sector and nurtured with financial and research assistance from NASA. Blended wing body designs are just that: aircraft in which the body and wing are not distinct structures, but shaped such that the body is a part of the wing and, thus, producing lift. A BWB design offers a number of possible advantages. When the fuselage is incorporated into the wing, the fuselage weighs less and creates less drag. With the wing taking up a larger area, wing loading may be reduced. Jack Northrop of the United States and Walter and Reimar Horton of Germany took this line of thinking to its logical conclusion in the 1940s by eliminating the fuselage entirely and building flying wings. A flying wing represents a design tradeoff, however, incurring difficulties with control, useful volume, ground handling, and wave drag at high speeds (unless it is an oblique flying wing). The BWB is a compromise between an all-wing structure and a segregated tube-and-wing structure.

NASA's involvement with BWB designs actually dated to the NACA period and the wind tunnel studies of simple, supersonic BWB designs. Supersonic BWB investigations continued into the 1960s with the search for viable supersonic transport configurations and have continued since.[22] It was not until the 1980s, however, that the idea of a subsonic BWB was taken seriously.

Tilt Nacelle V/STOL Aircraft" (Ames Research Center, Moffett Field, CA: NASA TM-86785, June 1986). On the wider body of V/STOL research, see W. P. Nelms and S. B. Anderson, "V/STOL Concepts in the United States—Past, Present and Future" (Ames Research Center, Moffett Field, CA: NASA TM-85938, April 1984).

21. Eric Lichtblau, "Leaders in House Block Earmarks to Corporations," *New York Times* (10 March 2010).

22. See, for example, George H. Holdaway and Jack A. Mellenthin, "Investigation at Mach Numbers of 0.20 to 3.50 of Blended Wing-Body Combinations of Sonic Design with Diamond, Delta, and Arrow Plan forms" (Ames Research Center, Moffett Field, CA: NASA TM-X-372, August 1960); William C. Sleeman, Jr., and A. Warner Robins, "Low-Speed Investigation of the Aerodynamic Characteristics of a Variable-Sweep Supersonic Transport Configuration Having a Blended Wing and Body" (Langley Research Center, Hampton, VA: NASA TM-X-619, 15 June 1967); and A. Warner Robins, Milton Lamb, and David S. Miller, "Aerodynamic Characteristics at Mach

It was at McDonnell Douglas in Long Beach, California, where engineers developed subsonic BWB designs for use in tactical military transports. NASA touched on this work when, from 1984 to 1985, Lewis Research Center contracted with the Long Beach group to study the application of propfans to their blended wing body tactical transport designs. Lewis was well into the Advanced Turboprop Project at the time.[23] In 1988, Robert H. Liebeck, a member of the McDonnell Douglas team that had worked on the Lewis contract, was inspired by Langley senior scientist Dennis M. Bushnell to consider the future of long-haul transportation. "Is there an aerodynamic renaissance for the long-haul transport?" Bushnell asked. Liebeck and his team took up the question and began considering the possibility of a large, long-range BWB for commercial service.[24]

Liebeck's team in Long Beach made an initial stab at a conceptual design that consisted of parallel tubular passenger compartments, so as to maintain pressurization at altitude, enclosed within a BWB shape. The initial calculations showed that a BWB design would, compared to a conventional design, be lighter, have a higher lift-to-drag ratio, and be more fuel-efficient. On this basis, NASA Langley contracted with Liebeck's group in 1993 to manage a feasibility study that included Langley, Lewis, and three universities. This was done under the Advanced Concepts for Aeronautics Program (ACP). The starting parameters were for an aircraft that could carry 800 people 7,000 nautical miles. Liebeck's group dispensed with the tubular arrangement of the first design in favor of a center-body box with a combined skin-pressure vessel made of a carbon fiber sandwich. The researchers mounted the engines in the trailing edge with intakes drawing air from the top of the BWB structure. They tested scale models in Langley's NTF and 14-by-22-foot tunnel, validating their CFD models. Stanford University built a 6 percent scale model in order to test low-speed characteristics. The study, completed in 1998, indicated that the BWB, compared to a conventional aircraft, would burn 27 percent less fuel, weigh 15 percent less, and have a 13 percent reduction in operating costs.[25]

Numbers of 1.5, 1.8, and 2.0 of a Blended Wing-Body Configuration With and Without Integral Canards" (Langley Research Center, Hampton, VA: NASA TP-1427, May 1979).

23. F. C. Newton, R. H. Liebeck, G. H. Mitchell, A. Mooiweer, M. M. Platte, T. L. Toogood, and R. A. Wright, Douglas Aircraft Company, "Multiple Application Propfan Study (MAPS) Advanced Tactical Transport" (Cleveland, OH: NASA CR-175003, 23 March 1989).
24. R. H. Liebeck, "Design of the Blended Wing Body Subsonic Transport," *Journal of Aircraft* 41, no. 1 (January–February 2004): 10–25.
25. Ibid., pp. 10–25; Robert H. Liebeck, Mark A. Page, and Blaine K. Rawdon, "Evolution of the Revolutionary Blended-Wing Body" (presentation at the "Transportation Beyond 2000:

It is important to note that this first stage of research was taking place at a time when the major aircraft manufacturers were casting about for designs in the "very large aircraft" (VLA) category. For the past three decades, the Boeing 747 had been the standard-bearer, a long-range aircraft that could accommodate approximately 420 persons in a typical three-class configuration. Although Boeing had revised the aircraft three times since the original, the aircraft no longer represented the most advanced aerodynamics, materials, or flight systems. McDonnell Douglas, prior to its decision to merge with Boeing, seriously considered and studied a large twin-deck configuration. The BWB was, obviously, a more radical design alternative, but it would have addressed the same market category. Airbus was especially interested in an aircraft as large as, or larger than, the 747 because Boeing was unopposed in that segment. A new VLA would round out Airbus's line and give the company, and Europe, new prestige. For a brief period of time, Boeing and Airbus conducted VLA studies jointly, the idea being that the market segment was too small for two competing aircraft. Eventually, Airbus decided to go it alone, designing and building the A380. Boeing pressed ahead with alternative studies, including the previously mentioned Sonic Cruiser, and the BWB research it inherited from McDonnell Douglas.

The ACP research was one of a number of investigations into the BWB at Langley. The Aeronautics Systems Analysis Division under Joseph Chambers was a strong proponent of expanded BWB investigations. The division's own design analysis indicated very attractive gains in efficiency, and those data would be reinforced by the ACP results. In 1996, Langley and industry partners approached NASA's leadership and made a pitch for a piloted, 26 percent scale experimental vehicle at a cost of $130 million. This was not approved.[26] Meanwhile, in Long Beach, the 3-year NASA-industry study prompted Boeing (which had merged with McDonnell Douglas) to sponsor its own BWB design study. This particular study began with an aircraft that had to fit within the standard large aircraft parking space (an 80-meter square box). The passenger count was 478 people in a three-class configuration. The researchers redesigned the shape using proprietary Boeing software and optimized it for easier manufacture. They also placed the engines on pods above the trailing edge, reducing the technical risks associated with resolving the flush-mounted inlet flows of the earlier design. The results of this study showed that the Boeing BWB had a 37 percent decrease in fuel consumption compared to a planned

---

Engineering Design for the Future" workshop, Langley Research Center, Hampton, VA, 26–28 September 1995); Chambers, *Innovation in Flight*, pp. 80–81.

26. Chambers, *Innovation in Flight*, pp. 79, 83.

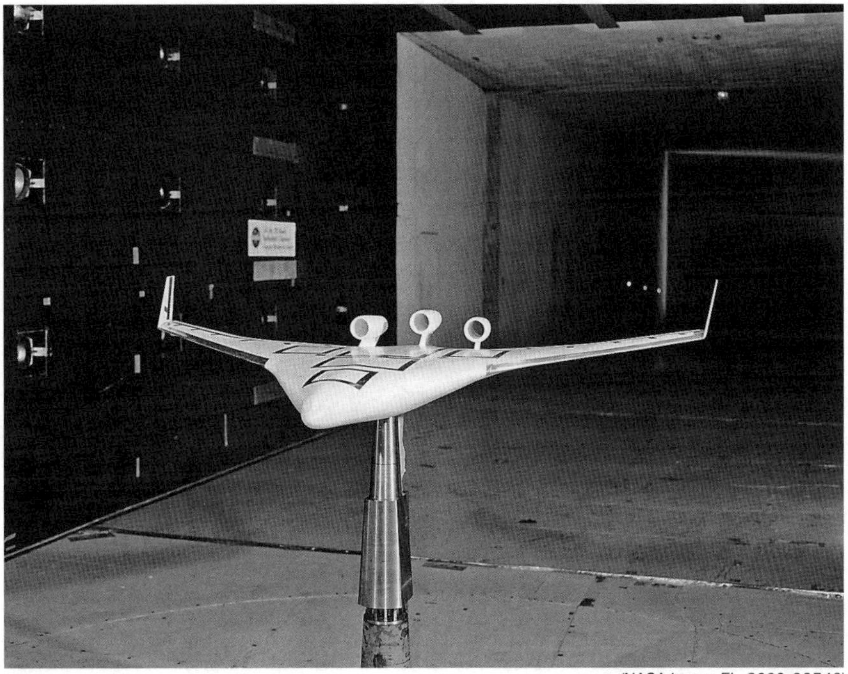

(NASA image EL-2000-00540)

Figure 7.1. Blended wing body model in the Langley 14-by-22-foot wind tunnel.

version of the Airbus A380. Additionally, the Boeing group had worked out a number of critical manufacturing details related to the complex center box pressure section.[27]

At Langley, there was a flurry of cross-disciplinary investigations on BWB questions. Since the vehicle represented such a departure from conventional design, Langley's scientists and engineers could not make quick assumptions about vehicle performance. They tested the aerodynamics, stability and control, spin, and noise characteristics. They analyzed BWB composite structural design, engine placement and design, and design methodologies. Toward the end of the decade, Langley's leadership made another attempt at finding top-level support for a BWB vehicle program. This time, however, the BWB was bundled into a new programmatic idea, Revolutionary Concepts (RevCon). RevCon was to be a series of 4-year design, fabrication, and test cycles covering a variety of revolutionary ideas. In 2000, the BWB became one of the first RevCon projects.[28]

---

27. Liebeck, "Design of the Blended Wing Body Subsonic Transport," pp. 10–25.

28. Chambers, *Innovation in Flight*, pp. 85–91.

(NASA image ED-07-0192-08)

Figure 7.2. X-48B scale model undergoing tests at Dryden in 2007.

The RevCon iteration of the BWB was a combined NASA-Boeing effort to produce and fly a low-speed, 14 percent scale, remotely piloted model. It was first called the BWB-LSV (low-speed vehicle) but later designated X-48A. The shape was based on Boeing's latest design iteration, the 478-person model. The X-48A was to have an impressive wingspan of 35 feet and be powered by three small turbojets. In-house construction began at Langley and was to be completed in 2002, with test flights scheduled in 2004. Unfortunately, the X-48A did not progress far before institutional and technical obstacles brought a halt to further work. The craft's designers were having difficulty with the vehicle's flight control system; while at Headquarters, aeronautics was, as noted earlier, rewriting its funding priorities.[29]

Langley was able to keep the program alive by shifting to a smaller 8.5 percent scale design and contracting fabrication of two models out to Cranfield Aerospace of the United Kingdom. With the Griffin reorganization, the BWB research became part of NASA's Fundamental Aeronautics subsonic fixed-wing

29. Ibid.; NASA Fact Sheet, "The Blended Wing Body: A Revolutionary Concept in Aircraft Design" (NASA FS-2001-04-24-LaRC, 2001).

effort. Langley researchers flew the first, unpowered model in the Full-Scale Tunnel. In July 2007, the powered model, the X-48B, flew from Dryden. Boeing and Langley were thus able to cross-validate flight-test, wind tunnel, and CFD data.[30] In addition to the flight tests, which understandably garnered the bulk of the public's attention, other researchers, often in conjunction with partners at universities and industry, examined a range of challenges associated with the BWB. Langley's structural researchers sought to find solutions to the noncylindrical fuselage, a shell structure that would be significantly different in manufacture and performance than traditional airframes.[31] Langley conducted studies into the acoustic advantages of the BWB. Both Langley and Glenn examined the placement, aerodynamics, and design of the BWB's propulsion.[32]

The Integrated Scramjet

In 2004, NASA flew an unpiloted scramjet-powered aircraft, first at Mach 6.8 and then at Mach 9.8. This was the culmination of more than four decades of scramjet research based at Langley. Langley had begun exploring the technology in the 1960s, when Lewis Research Center, the logical place for such work, was preoccupied with the space program. Langley's expertise and scramjet infrastructure grew such that by the turn of the century, it would have five test facilities at its disposal, allowing them to explore scramjet performance from Mach 3.5 to Mach 19.[33]

---

30. Tim Risch, Gary Cosentino, Christopher D. Regan, Michael Kisska, and Norman Princen, "X-48B Flight Test Progress Overview" (47th AIAA Aerospace Sciences Meeting, Orlando, FL, 5–8 January 2009); Chambers, *Innovation in Flight*, p. 85.
31. See, for example, V. Mukhopadhyay, J. Sobieszczanski-Sobieski, I. Kosaka, G. Quinn, and C. Charpentier, "Analysis, Design and Optimization of Non-Cylindrical Fuselage for Blended-Wing-Body Vehicle" (9th AIAA/ISSMO Symposium on Multidisciplinary Analysis and Optimization, Atlanta, GA, 4–6 September 2002).
32. Melissa B. Carter, Richard L. Campbell, Odis C. Pendergraft, Pamela J. Underwood, Douglas M. Friedman, and Leonel Serrano, "Designing and Testing a Blended Wing Body with Boundary Layer Ingestion Nacelles" (Langley Research Center, Hampton, VA: September 2006); Michael T. Tong, Scott M. Jones, William J. Haller, Robert F. Handschuh, "Engine Conceptual Design Studies for a Hybrid Wing Body Aircraft" (ASME Turbo Meeting, Orlando, FL, 8–12 June 2009).
33. The Langley scramjet facilities included the Direct-Connect Supersonic Combustion Test Facility, the Combustion-Heated Scramjet Test Facility, the Arc-Heated Scramjet Test Facility, the 8-Foot High Temperature Tunnel, and the Hypersonic Pulse Facility shock-expansion tube/tunnel. Ajay Kumar, J. Philip Drummon, Charles R. McClinton, and J. Larry Hunt, "Research in Hypersonic Airbreathing Propulsion at the NASA Langley Research Center" (Fifteenth International Symposium on Airbreathing Engines, Bangalore, India, September 2001 [ISABE–2001: Invited

Scramjets have long been a kind of "holy grail" of propulsion. They are jet engines, of sorts, but are more properly thought of as an alternative to rocket propulsion. Traditional jet engines operate at subsonic and supersonic speeds (up to the Mach 3 range in practice) and compress incoming air using fans (compressed air, as in a piston internal combustion engine, increases the oxidizer-to-fuel ratio). Ramjet engines do away with the compressor blades and use a vehicle's existing forward motion to compress the air. Obviously, a ramjet cannot operate from a dead stop and so must be paired with another initial propulsion system, such as a turbojet or rocket, or be able to convert from one type of engine to another in what is called a multi- or combined-cycle engine. The U.S. military conducted successful ramjet tests in the 1950s using the unpiloted X-7 vehicle series, reaching top speeds in excess of Mach 4. Ramjets saw operational deployment with the BOMARC and Talos missile systems.[34] The difference between ramjets and scramjets is the speed of the air inside the combustion chamber: in ramjets, the incoming air is reduced to subsonic speeds, whereas in scramjets, the air is moving at supersonic speeds. The ramjet is technically simpler, but the process of slowing the incoming air creates both pressure and heat; around Mach 6, the generated heat becomes a practical barrier. Scramjet combustion chambers operate at lower temperatures, but they are considerably more complex to design because of the challenge of reliably controlling the flow of supersonic air and igniting the fuel-air mixture. The theoretical advantage of the ramjet and scramjet, relative to rockets, is that they use atmospheric air as an oxidizer; rockets must carry their own supply. As a propulsion system, the scramjet has typically been envisioned as part of a combined-cycle engine to create an air-breathing orbital vehicle (e.g., by combining a rocket and scramjet) or as a hypersonic airplane (e.g., by combining a turbojet and scramjet).[35]

Though scramjets have long had proponents in the technical community, institutional support (from both NASA and the military) has waxed and waned over the decades. The history of scramjets is very much a tale of alluring

---

Lecture 4]). On the early roots of the Langley scramjet research, see Douglas L. Dwoyer, telephone interview by Robert G. Ferguson, 16 June 2007, NASA HRC.

34. Jenkins et al., *American X-Vehicles*; R. Stechman and R. Allen, "History of Ramjet Propulsion Development at the Marquardt Company—1944 to 1970" (41st AIAA/ASME/SAE/ASEE Joint Propulsion Conference and Exhibit, Tucson, AZ, 10–13 July 2005 [AIAA 2005–3538]); Paul J. Waltrup, Michael E. White, Frederick Zarlingo, and Edward S. Gravlin, "History of Ramjet and Scramjet Propulsion Development for U.S. Navy Missiles," *Johns Hopkins APL Technical Digest* 18, no. 2 (1997): 234–243.

35. Kumar et al., "Research in Hypersonic Airbreathing Propulsion," pp. 2–3.

theoretical possibilities, daunting technical challenges, and unstable political support. NASA's first scramjet program was the 1960s-era Hypersonic Research Engine (HRE), intended for testing on the X-15 rocket-plane. The HRE did not fly, but two models were tested in the Langley 8-Foot High Temperature Tunnel and the Plum Brook Hypersonic Test Facility.[36]

One of the early concerns of researchers was whether a scramjet would be able to overcome its own drag. Inspired by the results of the HRE experiments, Langley engineers began examining integrating the scramjet engine with the vehicle fuselage in order to reduce the overall drag. The vehicle's shape would form part of the inlet and nozzle, effectively making the fuselage functionally indistinguishable from the engine. This configuration became part of the DARPA Copper Canyon project and the subsequent multiagency National Aero-Space Plane program. (See chapter 5.) NASP, which ran from 1984 to 1995, was to employ a multicycle engine. This engine research was centered at Langley and saw the development of a concept demonstration engine (CDE), measuring 10 by 16 by 142 inches. Langley engineers ran the CDE successfully in the 8-Foot High Temperature Tunnel at Mach 6.8.[37]

As NASP began drawing down, Langley engineers sought to continue scramjet research by proposing a very modest development and test program.[38] Whereas the NASP program sought to build an actual, full-size vehicle, the Langley proposal envisioned a remotely piloted scale model, carried aloft by a B-52 and boosted to the proper speed and altitude by a commercial rocket. Named the X-43A Hyper-X, the craft would do what had never been done before: flight-test a fuselage-integrated scramjet and validate design models. The proposal, approved by NASA Headquarters in 1996, initially called for four tests: one at Mach 5, one at Mach 7, and two at Mach 10. A year later, the plans included only three flights: two at Mach 7 and one at Mach 10.[39]

The X-43A vehicle was, relative to many of the scramjet plans that preceded it, exceedingly small, measuring only 12 feet long and 5 feet wide. It incorporated the integrated fuselage-engine design that Langley engineers

---

36. Ibid., pp. 1–2.

37. Ibid., p. 3; Larry Schweikart, *The Quest for the Orbital Jet: The National Aero-Space Plane Program (1983–1995)*, vol. 3 of *The Hypersonic Revolution: Case Studies in the History of Hypersonic Technology* (Washington, DC: Air Force History and Museums Program, 1998).

38. There were a variety of scramjet proposals over the years, both at NASA and at the U.S. Air Force. These are nicely detailed in Curtis Peebles, "The X-43A Flight Research Program: Lessons Learned on the Road to Mach 10" (unpublished manuscript, NASA Dryden Flight Research Center, 2007).

39. Peebles, "The X-43A Flight Research Program," pp. 46–52, 66.

had advocated 20 years earlier and took the shape of a lifting body. Its only appendages were two horizontal and two vertical stabilizers. The winning bid to build three vehicles, which were to be flown once and not recovered, went to MicroCraft of Tullahoma, Tennessee, over the much larger and more experienced rival, McDonnell Douglas. McDonnell Douglas, in fact, had performed earlier scramjet engine research under contract. The DF-9 project, as it was known, was a progenitor of the Langley Hyper-X engine that would fly on the X-43A.[40]

The X-43A faced many design challenges. At such high velocities, the craft would be subject to brutal conditions, both in terms of heating (caused by aerodynamic friction) and control (a situation exacerbated by a rocket boost-separation sequence). The vehicle would require special heat-resistant materials, as well as methods for cooling internal systems. Because the engine was integrated into the fuselage, there was close coupling in the design of different systems (i.e., changes in one system had complex knock-on effects for other systems). The X-43A faced significant engineering challenges in the design and operation of the rocket boost system. Engineers chose to use a commercial rocket, the Orbital Sciences Pegasus booster. The Pegasus, a winged rocket air-dropped from a B-52, would follow a shallow ascent trajectory designed to place the X-43A within a narrow speed and altitude window, at which point the X-43A would separate from the booster and, if all went as planned, open its scramjet engine doors, add hydrogen, and ignite the fuel-air mixture. The integration of all these different flight elements fell on teams at Langley and Dryden, with the latter charged with integrating and testing the various systems before the flight test.[41]

Unfortunately, the first flight in June 2001 went awry as the X-43A/booster combination spun out of control shortly after launch. Range operators detonated the booster, the scramjet never having had the opportunity to run. A subsequent investigation determined that there were modeling errors in the design of the control system. When the booster behaved in an unexpected manner after air-drop, the control system was unable to effectively respond.[42]

Nearly three years later, after recalculating aerodynamic loads and making changes to the control system, the second of the three vehicles was ready for a new attack on the Mach 7 target. On 27 March 2004, after a successful air-drop, the X-43A separated from the Pegasus booster at Mach 6.95. With drag

40. Ibid., pp. 56, 98–100, 112–115. MicroCraft was subsequently purchased by Alliant Techsystems, Inc. (ATK) of Minnesota.

41. Ibid., chaps. 1–2.

42. Ibid., pp. 168–187.

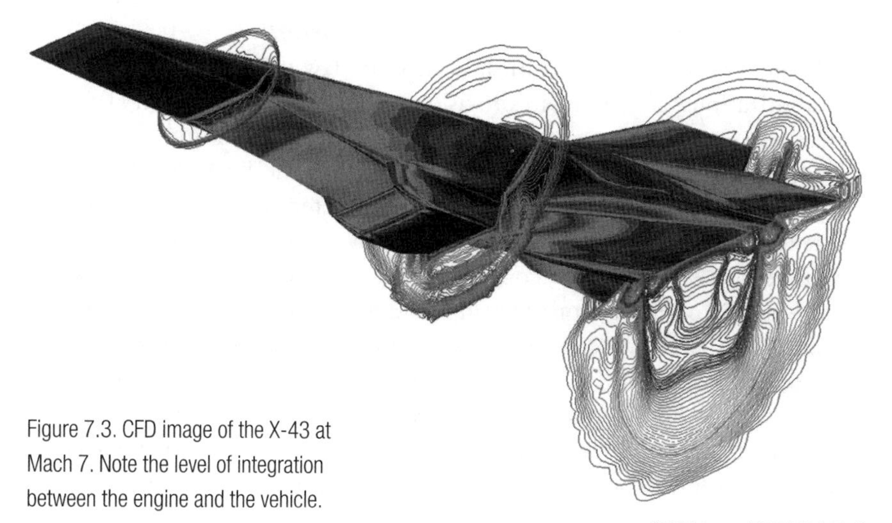

Figure 7.3. CFD image of the X-43 at Mach 7. Note the level of integration between the engine and the vehicle.

(NASA image ED97-43968-1)

slowing the vehicle down precipitously, the X-43 only had a few seconds to stabilize, open the scramjet's inlet, and inject and ignite gaseous hydrogen. The burn lasted 10 seconds, and the vehicle reached a top speed of Mach 6.83. It was a successful test, with the vehicle generating more thrust than drag. The following November, the Hyper-X program launched its last vehicle. The craft separated from the Pegasus at Mach 9.74 and nearly 110,000 feet; after ignition, it reached a speed of Mach 9.68.[43]

As the X-43A designation implies, there were follow-on designs (i.e., X-43B, X-43C) that were meant to test combined-cycle engines, as well as different fuels. The program came to an abrupt end shortly after the November flight, however. Indeed, researchers had expected funding to continue until at least March 2005, but it ran out within weeks, before the team even had time to analyze and publish their results.[44] While NASA's Aeronautics Directorate maintained a hypersonics project within its Fundamental Aeronautics Program, the Air Force Research Laboratory's Hypersonic Technology Office took effective lead of scramjet vehicles and funded the development of a Boeing–Pratt & Whitney hydrocarbon-fueled scramjet designed to fly at Mach 6–7 (designated the X-51). NASA remained an important participant in this research.[45]

43. Ibid., pp. 231–235, 273.

44. Ibid., pp. 281–282.

45. Richard R. Kazmar, "Airbreathing Hypersonic Propulsion at Pratt & Whitney—Overview" (AIAA/CIRA 13th International Space Planes and Hypersonics Systems and Technology Conference,

Next Generation Air Transportation System

In December 2003, President Bush signed into law the Century of Aviation Reauthorization Act. Among its provisions was the implementation of a Next Generation Air Transportation System (NGATS) to be coordinated out of a Joint Planning and Development Office within the FAA. NGATS goals were to design a new air traffic management system that increased system efficiency and safety. NGATS was to take advantage of new technological developments, especially new computing and communications technologies, while building on the nation's existing (and much more antiquated) air traffic control system. The senior policy committee of the Joint Office included representatives from the FAA, NASA, DOD, Homeland Security, Commerce, and the OSTP. Additionally, NGATS was to make extensive use of input from airlines and equipment providers. All of this was very much a reflection of the difficulties encountered in past attempts to upgrade aspects of the air transportation system.[46]

Since 1958, the nation's air traffic control (ATC) system had been under the Federal Aviation Administration (originally named the Federal Aviation Agency). The system of airports and airways relied in large part on control methodologies that dated to the World War II era, namely ground-based controllers equipped with radar who provided clearance and assigned routes, via radio, to pilots. The pilots, for their part, navigated via a system of radio beacons and inertial reference devices and were able to descend and land by instruments tuned to the airport's instrument landing system.[47] Airlines added

---

Capua, Italy, 16–20 May 2005 [AIAA 2005–3256]). One indicator of the technology's potential military utility is that the X-43A engine performance data were always classified.

46. *Vision 100—Century of Aviation Reauthorization Act*, H.R. 2115, 108th Cong., passed 12 December 2003. See also FAA, Joint Planning and Development Office, "Next Generation Air Transportation System Integrated Plan," December 2004.

47. The addition of radar, strictly speaking, was considered to be the second generation of air traffic control. The development of ATC radar dated to World War II but was implemented widely in civil aviation in the 1950s. The third-generation ATC system, implemented in the late 1960s and early 1970s, included small amounts of automation for the controller, as well as the automatic broadcast of aircraft data to ATC centers. On the early design of air traffic control, see Radio Technical Commission for Aeronautics, Special Committee SC31, "Air Traffic Control" (Paper 27-48/DO-12, 12 May 1943); U.S. Department of Transportation, "Report of the DOT Air Traffic Control Advisory Committee," vol. 1, December 1969; Federal Aviation Agency, System Research and Development Service, "Design for the National Airspace Utilization System: Summary of First Edition," September 1962; and Erik M. Conway, *Blind Landings: Low-Visibility Operations in American Aviation, 1918–1958* (Baltimore: Johns Hopkins University Press, 2006).

their own managerial layer with operations centers that dispatched flights and filed flight plans. The control and utilization of the airspace was (and remains) a complex human-machine system with responsibility divided among pilots, controllers, and airlines using matched ground-aircraft technical systems.[48]

Airspace crowding emerged as a public issue in the 1960s, with the chief problem being congestion in the terminal area (i.e., aircraft landings). Recognizing that ATC was becoming overburdened, the newly minted Department of Transportation formed the Air Traffic Control Advisory Committee in 1968. The committee, numbering nearly 150 persons, was charged with recommending an ATC system for the 1980s and beyond. The committee's report, known familiarly to those in the field as the Alexander Report, argued for increased automatic communication and control methods.[49] NASA had two high-profile programs aimed, directly and indirectly, at solving the problem of congestion; both addressed the issue from the vehicle side. At Ames, the short takeoff and landing (STOL) program sought to develop aircraft that could operate out of smaller fields, thus alleviating pressure on conventional runways. Researchers at Langley initiated the Terminal Configured Vehicle (TCV) Program, which sought new avionics and flight procedures that would increase safety and productivity.[50] The dark horse, however, was a less well-known effort led by Dr. Heinz Erzberger at Ames. Erzberger's work, ultimately, would be the seed for NASA's airspace systems focus three decades later.

Erzberger, who was familiar with the Alexander Report's conclusions, decided to approach the problem as a mathematical puzzle to find, analytically, a way to automate the greatest bottleneck in the ATC system. Erzberger's solution was not a piece of equipment, but an algorithm, a way of calculating where an aircraft should be in its approach as a function of time. Erzberger termed this "four-dimensional guidance," and the feature that distinguished it from subsequent competing efforts was its time-based sequencing (the fourth dimension). Initially conceived as a program that could be run from a computer in an aircraft, it also could form part of an automated ATC system, where target landing slots drove the algorithm's time parameters.[51] For the 1970s, however, Ames researchers quietly tested the 4D landing trajectory algorithm under the cover of Ames's STOL research (called the STOLAND project); Erzberger and

---

48. Ibid.
49. DOT, "Report of the DOT Air Traffic Control Advisory Committee," vol. 1, December 1969, p. 59.
50. "Program Plan for Terminal Configured Vehicle Program" (Langley Research Center, Hampton, VA: NASA TM-108227, 1 December 1973).
51. Heinz Erzberger and Homer Q. Lee, "Terminal-Area Guidance Algorithms for Automated Air-Traffic Control" (NASA TN D-6773, April 1972).

his colleagues were fearful of treading on the FAA's ATC-oriented research, as well as Langley's purview over conventional takeoff and landing.[52]

For a period, Erzberger left 4D work to create algorithms for flight management systems (on-board computers that, for example, optimize flight profiles to minimize fuel consumption). He returned to the work in the 1980s, this time arguing for the direct use of 4D algorithms for software in air traffic control.[53] To test and improve this software, however, Erzberger reasoned that he needed live data from one of the FAA's terminal radar approach controls (TRACONs). This would not be easy to acquire, especially since Erzberger was encroaching on the FAA's territory. Indeed, this took place at about the same time that the FAA was carrying out a raft of system upgrades. Two of the most high-profile, and most expensive, included the Advanced Airspace System (AAS) and the Microwave Landing System (MLS). AAS was a multibillion-dollar program begun in 1983 to replace many of the ATC computers and, in the process, consolidate facilities and automate many controller functions.[54] MLS was meant to replace the less accurate, World War II–era ILS system. Initial work began in the 1970s, and the FAA anticipated installation at all of the country's international airports by 1998.[55] Additionally, the FAA had been supporting a significant research effort into automated traffic control. From 1976 to 1990, the MITRE Corporation had, under contract, been developing an Automated En Route Air Traffic Control (AERA) program.[56]

---

52. Heinz Erzberger, telephone interview by Robert Ferguson, 27 February 2008; Homer Q. Lee, Frank Neuman, and Gordon H. Hardy, "4D Area Navigation System Description and Flight Test Results" (NASA TN-D-7874, August 1975).
53. Erzberger interview, 27 February 2008; Heinz Erzberger and Leonard Tobias, "A Time-Based Concept for Terminal-Area Traffic Management" (NASA TM-88243, April 1986).
54. The original cost was $2.5 billion in 1983 but had risen to $4.8 billion when contracts were signed in 1988 and continued to grow to an estimated $5.1 billion by 1993. U.S. Government Accounting Office, "Advanced Automation System: Implications of Problems and Recent Changes," testimony of Allen Li, Associate Director, Transportation Issues, Resources, Community, and Economic Development Division, before the Subcommittee on Aviation Committee on Public Works and Transportation, House of Representatives, GAO/T-RCED-94-188, 13 April 1994.
55. John H. Cushman, Jr., "Aircraft Landing System of Future Is Inaugurated," *New York Times* (7 April 1989).
56. Lawrence Goldmuntz, John T. Kefaliotis, D. Weathers, Louis A. Kleiman, Richard A. Rucker, and Leonard Schuchman, "The AERA Concept" (U.S. Department of Transportation, FAA, Office of Systems Engineering Management, Fm-EM-81B3, 24 March 1981); Heinz Erzberger, "The

Erzberger finally received approval for the live FAA data feeds in the late 1980s and began developing the Center-TRACON Automated System (CTAS). CTAS was as much a research methodology as it was a prototype. Researchers performed simulations at Ames in an ATC simulator that integrated data from the TRACONs at Denver and Forth Worth, as well as piloted simulators at Ames and Langley. The tools were more sophisticated versions of the algorithms described above. One tool, the Traffic Management Advisor, assigned runways and landing sequences and times. Another, the Descent Advisor, told aircraft cruising at altitude when to begin their descents to the landing pattern. And lastly, the Final Approach Spacing Tool did just that, telling controllers how to space aircraft for final approach. The CTAS program began field-testing this software suite at the Denver TRACON in the fall of 1995 and at the Fort Worth TRACON in the summer of 1996. Interestingly, the tests at Fort Worth were so well received that the controllers, the National Air Traffic Controller Association, and the Air Transportation Association asked that the FAA and NASA keep the prototype system in place until the FAA actually upgraded the system with the same capability.[57]

The FAA's own efforts at modernization, however, were not going so well. The Advanced Airspace System was well over budget and behind schedule. The 1988 cost estimate of $4.3 billion had grown to $5.9 billion by 1993. The GAO testified to Congress that the FAA had been overly ambitious in its plans. Even the FAA's own AAS Task Force estimated that if the program were left to run its course, it would end up costing $6.5 to $7.3 billion, with much of the increase coming from software development. Ultimately, the FAA greatly reduced the scope of the AAS implementation, curtailing the more advanced automation functions that it was supposed to deliver. By 1996, the AAS effort was considered a fiasco, with about half a billion dollars spent on software that would not be used.[58] For different reasons, the Microwave Landing System

---

Automated Airspace Concept" (4th USA/Europe Air Traffic Management R&D Seminar, Santa Fe, NM, 3–7 December 2001).

57. Dallas G. Denery and Heinz Erzberger, "The Center-TRACON Automation System: Simulation and Field Testing" (NASA TM-110366, August 1995); S. M. Green and R. Vivona, "Field Evaluation of Descent Advisor Trajectory Prediction Accuracy" (AIAA Guidance, Navigation, and Control Conference, July 1996 [AIAA 96-3764]); H. N. Swenson, T. Hoang, S. Engelland, D. Vincent, T. Sanders, B. Sanford, and K. Heere, "Design and Operational Evaluation of the Traffic Management Advisor at the Fort Worth Air Route Traffic Control Center" (1st USA/Europe Air Traffic Management R&D Seminar, Saclay, France, June 1997).

58. Matthew L. Wald, "Flight to Nowhere: A Special Report; Ambitious Update of Air Navigation Becomes a Fiasco," *New York Times* (29 January 1996); U.S. GAO, "Advanced Automation System."

also never saw full implementation, and by 1994, the FAA had canceled the two-and-a-half-decadelong project.[59] MLS, unlike AAS, worked, but it had never earned the full support of the airlines, who were not keen to install new navigational equipment costing $250,000 per aircraft. The knockout blow came from Global Positioning System (GPS) technology developed by the Department of Defense. GPS had strong advantages over MLS: it was already deployed by the time MLS was beginning operational use, and GPS provided navigational data throughout a flight, not just during the landing. Airlines willingly installed GPS equipment to help increase the accuracy of aircraft inertial navigation systems. Interestingly, although NASA had assisted the FAA in testing and demonstrating MLS systems, NASA also flight-tested early GPS equipment, helping to show that this was a potential alternative to MLS.[60]

Against this background, NASA's very-small-scale foray into ATC was something of a miracle. CTAS worked and, up to that point, the FAA's major contribution had been TRACON data feeds. Following the initial set of CTAS software tools, NASA's researchers expanded their reach, designing programs that assisted in en route and departure control. Such programs continually monitored and optimized aircraft routes in order to increase efficiency and safety. These too were tested in the field using the CTAS methodology.[61] CTAS design/test methodology would lead to more advanced airspace simulation systems designed specifically to test new air traffic management technologies.[62]

Through the 1990s, this work was carried out under the umbrella of the Advanced Air Transportation Technologies (AATT) program.[63] By the end of the 1990s, NASA's management had grown increasingly fond of the ATC research, elevating it higher and higher within the Agency's mission focus. It

59. Martin Tolchin, "U.S. Cuts Back $7 Billion Plan for Air Traffic," *New York Times* (4 June 1994).

60. Lane Wallace, *Airborne Trailblazer: Two Decades with NASA Langley's 737 Flying Laboratory* (Washington, DC: NASA SP-4216, 1994), chap. 4.

61. B. D. McNally, S. Engelland, R. Bach, W. Chan, C. Brasil, C. Gong, J. Frey, and D. Vincent, "Operational Evaluation of the Direct-To Controller Tool" (4th USA/Europe Air Traffic Management R&D Seminar, Santa Fe, NM, 3–7 December 2001); S. J. Landry, T. Farley, and T. Hoang, "Expanding the Use of Time-Based Metering: Multi-Center Traffic Management Advisor" (proceedings of the 6th USA/Europe ATM 2005 R&D Seminar, Baltimore, MD, 27–30 June 2005).

62. See, for example, S. Zelinski, "Validating the Airspace Concept Evaluation System Using Real World Data" (AIAA Modeling and Simulation Technologies Conference and Exhibit, San Francisco, CA, August 2005 [AIAA 2005–6491]).

63. David R. Schleicher, George J. Couluris, and John A. Sorensen, "National Airspace System Operational Concept—AATT Products Mapping Analysis" (Berkeley, CA: National Center of Excellence for Aviation Operations Research, Report No. RR-98-1, March 1998).

did not hurt that this technology was a public good and so did not run afoul of the more ideological objections to federally funded research. By 2005, as noted above, airspace systems research had become one of the four main areas of the Aeronautics Research Mission Directorate, an amazing rise for an effort that began as an algorithm that could have been written on a chalkboard three decades earlier.

Still, by the turn of the century, the nation's airspace remained a hodgepodge of old and new technology and practices. The difficulty in bringing change to the air traffic control system was very much a reflection of the technical and organizational complexity of the system. Erzberger and coauthor Leonard Tobias nicely captured the pitfalls awaiting ATC modernization in a 1986 paper (well before the turmoil of the FAA's AAS):

> At first, problems in ATC automation often do not appear to be more difficult than typical aircraft guidance and control problems that have been successfully solved. But then, after some promising initial successes, unforeseen problems surface and reach unmanageable complexity as more and more practical constraints are included, leading to the eventual abandonment of the effort.[64]

As antiquated as the FAA's air traffic control system was, it was also extremely flexible. Human controllers might not be as efficient as computers, but they could recognize and handle a large variety of nonstandard conditions. Herein lay an important advantage in the CTAS approach; using live data feeds, the NASA researchers gained an intimate picture of the complexity of ATC operations. Each lesson they learned became a new twist in the CTAS algorithms.[65]

As of the writing of this book, NGATS is very much a work in progress. Designed to reduce the development risks that befell the AAS program, NGATS has a long period of evaluation and testing, with constant input from the stakeholders. The research and validation phase of numerous subprograms is to continue to 2020, with staged, concurrent transition phases leading to implementation by the year 2025. The ultimate goal is to provide a threefold increase in airspace use. An example of the institutional and technical revolution imagined by NGATS is the switch from sector-based control to trajectory-based control. As noted above, pilots receive clearances from controllers, but these controllers are responsible for only a portion of the airspace

64. Heinz Erzberger and Leonard Tobias, "A Time-Based Concept for Terminal-Area Traffic Management" (NASA TM-88243, April 1986).
65. Erzberger interview, 27 February 2008.

(a sector). Through a single flight, airline pilots will receive clearances from 10 to 20 different controllers, each operating his or her own sector. Not only is this unwieldy from the pilot's perspective, but the system is poorly structured to take advantage of the efficiencies that a computer-managed airspace can provide. Sectors, fixed routes, and waypoints, fundamental elements of the traditional human-based system, are unnecessary in a trajectory-based system. The computer, for example, has the ability to do away with fixed routes and waypoints and adjust the aircraft's trajectory as conditions merit, a method conceptually known as a dynamic airspace.[66] Still, any dramatic changes to air traffic control will have to be done, as Erzberger came to realize, in conjunction with the existing ATC system and its human controllers. There is no such thing as replacing human operations without first integrating the new procedures *into* human operations.

Conclusion

Given the speed with which Washington and NASA's leadership promulgated Agency goals and policies, it is tempting to argue that this is only so much gloss, that the underlying research persisted regardless. Certainly, considering the lengthy development periods of the three conceptual developments examined in this chapter, R&D seems relatively immune to political fortunes. Scramjets, blended wing bodies, and 4D ATC algorithms survived multiple decades. Yet, under closer examination, it is clear that Agency realignments did buffet and threaten work at the Centers. The BWB case illustrates how Center researchers responded to (indeed, embraced) calls for greater customer outreach and revolutionary technologies. Langley's RevCon seemed to combine aspects of the 1990s faster, better, cheaper approach with an emphasis on revolutionary thinking. BWB advocates assumed the language of revolution and jockeyed to keep programs funded. Their efforts did not lead to consistent political support, and time after time, researchers scaled down the scope of their research to fit smaller and smaller budgets. The history of scramjet research, played out over an even longer period, mirrored the BWB's battles over shifting top-level political and technical goals while also finding tenuous success in an abbreviated, small-scale test program.

66. P. Kopardekar, K. Bilimoria, and B. Sridhar, "Initial Concepts for Dynamic Airspace Configuration" (7th AIAA Aviation Technology, Integration, and Operations Conference, Belfast, Northern Ireland, 18–20 September 2007 [AIAA 2007–7763]); Harry Swenson, Richard Barhydt, and Michael Landis, "Next Generation Air Transportation System (NGATS) Air Traffic Management (ATM)—Airspace Project" (Ames Research Center, Moffett Field, CA: NASA Technical Report, June 2006).

Aeronautics' realignments reflected substantial uncertainties within Headquarters about optimal funding and R&D management, but it was also part of a trend that saw programmatic control drifting away from the Centers. There were efforts from Headquarters to reverse this (e.g., through the establishment of lead Centers), but it was still the case that battles over program definition were being fought in Washington, DC. The growth of congressional earmarks was but one manifestation of the increased politicization of NASA's budget. As NASA enters its sixth decade, the question for aeronautics is whether the latest reorganization will provide researchers with long-term stability for laboratory-led research.

Bucking this trend was the good fortune enjoyed by air traffic control research. To some extent, this growing area benefited from astute technical choices made decades ago; 4D algorithms provided an elegant basis for practical ATC tools. ATC research also benefited from airspace congestion that was only getting worse, but the key innovation that broke down barriers between the FAA and NASA was the CTAS methodology. With a modest investment (relative to earlier FAA-sponsored R&D), the Ames ATC team created an unobtrusive method for testing and refining its 4D algorithms. One might think of CTAS as a kind of wind tunnel for ATC research, though what the live TRACON data feeds really provided was clinical data. Ames researchers learned how ATC was truly done. Ames's 4D algorithms might have ended up as just another failure in the history of ATC automation, but instead, field trials exceeded expectations. With this, a dark horse area of research gained political stature within NASA and legitimized NASA's participation in what was once the sole purview of the FAA.

One last observation concerns the origins of conceptual developments. All three of the concepts in this chapter were, one could argue, outsider developments. Langley, not NASA's propulsion laboratory, was the center for scramjet research. The 4D ATC algorithms emerged at NASA (and were initially buried within STOL research) rather than at the FAA and its contractors. The subsonic blended wing body emerged with a contractor and was taken up by NASA's laboratories. Contemplated with other developments highlighted in this book, such emerging technologies by outsiders are not unusual. In light of the emphasis paid to Agency and programmatic goals, to budgetary realignments and organizational juggling, to stated research strengths and focal points, it is clear that innovative thinking defied institutional attempts to predict, measure, and harness it.

Chapter 8: Conclusion

Upon reaching the conclusion of a survey on over 50 years of aeronautical research, one might expect a tally of the Agency's accomplishments. This is problematic, if not unsatisfying. The economy of this narrative has forced us to jump from one project to the next, leaping past stretches of research, past individuals and communities, past accomplishments that are as worthy as the ones I have chosen to include. To reduce this narrative's assemblage further into a kind of bumper-sticker encomium undermines our grasp of history's complexity and texture. Instead, it is more fruitful to reassess some of the major topics posed at the start: the impact of the space program, the evolution of the laboratories, political control and the resiliency of the scientists and engineers, and, lastly, why NASA does aeronautics research at all.

On the Impact of the Space Program

One of the questions that this book first posed was what kind of impact the space program had upon aeronautics research. The most basic question is whether aeronautics would have been better off going it alone, rather than becoming subsumed into a much larger space-centric agency. In all likelihood, the narrative suggests, in the absence of Sputnik or a combined space agency, a national aeronautics research program would have persisted, at least to the 1990s. The Cold War was sufficient motivation for the nation to continue funding a broad, multicenter aeronautical research program. The fact that the Soviet Union developed its own research programs guaranteed an American response. Whether aeronautics would have survived is a different question from whether it would have been better off. This, however, is a hypothetical question that the narrative cannot answer. What we can identify are the qualitative changes wrought by partnership with the space program.

The first is the opportunity for aeronautics researchers to work on space-related topics and transfer to the space program. For many scientists, engineers, and managers, especially in the nascent days of NASA, the space program greatly energized the Centers. True, there were individuals who were threatened by the transition, who worried that aeronautics would be lost, but the space program took along many prominent aeronautical researchers, especially at Langley, which was the early institutional base for space planning. Even those researchers who did not make the transition had the freedom to explore areas that bridged the two worlds. There were continuing research synergies in a number of fields, such as lifting-body vehicles, hypersonic flight, scramjets, and digital flight control. In a combined agency, support for such efforts naturally would have been easier to gain and communication more fluid among the various aeronautics and space teams.

The second broad development was the administrative influence from the space program. Here one must be careful not to ascribe to the space program institutional changes that were occurring across the federal government (e.g., the 1993 Government Performance Results Act). Increased budgetary control and, with that, increased centralization over planning and management were long-term trends in the evolution of the federal government. Where the space program had a more salient impact was on priority and strategy. The human space program has always been the center of NASA's mission, and even though the Agency has many other programs, human space is really why the Agency exists. It not only consumes the lion's share of the budget, it is also the spiritual core of the Agency. It is not going out on a limb to say that the NASA Administrator's primary job has always been the planning and execution of the human space program. In this light, it is not surprising that aeronautics has not always enjoyed the close attention of NASA's leadership. This is a structural tension that, though addressed most recently in the Griffin reorganization, will continue to exist so long as the human space program requires the bulk of the Agency's financial and managerial resources. There is a flip side to the low priority assigned to aeronautics. For at least the first decade, there was fairly little meddling from Headquarters in the structure and content of the research at the Centers (save that Lewis transitioned to space-based propulsion problems). Longtime NASA researchers fondly remember those days as a period of greater local control over funding decisions.

As for strategy, the space program created a strong precedent for mission-oriented programs that, at least at the organizational level, pulled the Agency away from unquestioned support for fundamental research. Recall that at the beginning of NASA's existence, aeronautics was supposed to be part of a core research program distinguished by the fact that it was not performing missions. By 1986 and the Space Shuttle Challenger disaster, pundits and NASA leadership alike were wondering whether NASA had lost its way and musing that perhaps it needed to be reenergized with bold, new missions. Aeronautics was not immune to these leadership shifts and so, by the 1990s, had its own visionary goals more befitting a massively financed and centrally executed space program than the kind of laboratory-based work to which it was accustomed. Here again, the most recent reorganization sought to blunt these trends.

On Laboratory Evolution

At the beginning of this book, I mentioned the question of technological progress and the tension between technology as a social process and technology as a kind of logical unfolding predicated on available knowledge, resources, and physical constraints. It is useful to return to this question in thinking about NASA's scientific and technological narrative. I illuminated one side of this argument in the discussion of digital electronics as an enabling technology. Digital fly-by-wire was a stepping-stone for a new era in unstable experimental aircraft. The other side of the argument is ably illuminated by the trajectories of the various Centers. Far from following a logical script, the Centers grew in ways that could not have been anticipated or centrally coordinated.

The arrangement of the four Centers was largely a product of World War II, created to increase laboratory capacity, better serve far-flung industrial and military patrons, and, in the case of the Muroc facility, take advantage of prime testing conditions. Over the long term, these communities formed distinct identities and approaches. What at times has been considered an Achilles heel of having multiple Centers, i.e., the duplication of research infrastructure, actually contributed to alternative pathways. In the absence of this socially constructed arrangement, we would likely be telling a very different narrative. Cases from Ames and Dryden best illustrate the divergence in research trajectories.

Even prior to the NASA era, Ames and Langley had taken on different characteristics, very much the result of the influence of their lead scientists and engineers. In the 1960s and 1970s, Ames leaders made explicit moves to differentiate the Center's work from Langley's. Ames's infrastructure mirrored many of the capabilities of Langley, and with Langley as the historic Center for aeronautics, Ames appeared particularly vulnerable to budget cuts. Ames sought out alternative projects and methodologies; the most obvious of these was its wager on computational fluid dynamics. Competing head-to-head with Langley for wind tunnel research was not only a zero-sum game, it was one that Langley was likely to win over the long term. Ames chose to pursue CFD not only because the time seemed right for the technology, but also because it was a potential institutional lifesaver. Langley also pursued CFD, but it did not attach itself to the same vision of supercomputing as Ames. In the end, this proved a fruitful rivalry.

Dryden illustrates a more subtle form of differentiation. On its face, Dryden was created as a simple flight-test facility. As such, Dryden was originally an adjunct to Langley and Ames, the last link in a chain of experimentation. Instead, Dryden became a research facility in its own right, and rather than competing with Ames and Langley by duplicating infrastructure, its researchers considered how they could turn a dry-lakebed flight-test facility into a primary research tool. Dryden personnel began dreaming up their own research projects. Lacking wind tunnels, they sought help from their peers at Langley and Ames. Lacking deep pockets, they explored cost-effective methods for getting their ideas aloft, notably scale models and remotely piloted vehicles. The lifting-body program of the 1960s is remarkable both for its technical achievements and as a project bootstrapped from hand-built glider models. The Dryden approach was frugal and had the advantage of returning actual flight data.

Having multiple research centers has been one of NASA's underappreciated strengths. Despite recurring efforts to rein in and coordinate Center operations, the duplication of infrastructure (and the occasional threat to eliminate it) has had a serendipitous effect on output. Differentiation became a matter of survival. More fundamentally, competition among multiple Centers gave rise to alternative methodologies. It was not that Centers decided to divide up the research pie (which they certainly did do), but that they spurned conventional wisdom in order to establish new niches.

Political Control and Resilient Scientists and Engineers

Looking more closely at the content of laboratory research over the years, we see an amazing resiliency in communities and fields of research. In some cases, their persistence defies aviation's technological trends, not to mention uncertain policy support and periods of on-again, off-again funding. Research into supersonic commercial aircraft, for example, has continued doggedly in spite of dim market prospects that have been evident since the United States canceled its own SST in 1971. There is in any field, of course, an instinct for self-preservation, but the complexity of aerodynamic phenomena also ensures that there is no shortage of questions that might be asked. So even though the aerodynamic understanding of current subsonic commercial aircraft has reached a high level of maturity, and even though the market has reached closure on these designs (i.e., that they should be conventional tube-and-wing structures), there is still plenty of unexplored territory in unconventional designs like the blended wing body or oblique wings. Regardless of these concepts' ultimate fate, such investigations inform our fundamental understanding of physical phenomena and technical performance.

To outsiders, the enthusiasm for the unproven and out-of-fashion may seem an example of waste, fraud, and abuse within the federal government. Should not NASA be pursuing ideas that have realizable, practical ends, militarily or commercially? Would not the Agency be more efficient if it could carefully choose which ideas to support, weeding out the impractical, the fanciful, and the poor performers? Put more succinctly, shouldn't research funding mirror sociotechnological cycles? The historical record suggests that this is not such an easy proposition. There is, first of all, the problem of trying to predict with any reasonable level of accuracy where society and its technologies will be a decade or more in the future. Second, there is the problem of creating and implementing roadmaps. The strategy gyrations of the late 1990s and early 2000s are testament to the difficulty of creating meaningful goals within a highly politicized context. Third, there is the problem that research projects are not disembodied commodities that can be chosen at will. NASA is ultimately supporting what is supposed to be a community of top-notch researchers, and even the best of these will have ideas that do not pan out. Richard Whitcomb, one of the Agency's most famous aerodynamicists, proposed and refined the idea of area-ruled fuselages for commercial transports. However, at the speeds that commercial aircraft operate, this was both unnecessary and impractical (given the manufacturing costs of fabricating non-uniform fuselage sections). Who in the 1960s would have been sufficiently prescient to tell Whitcomb which of his projects he should pursue? Furthermore, fruitful ideas do not necessarily materialize when and where the Agency chooses. Heinz Erzberger's work on air traffic control algorithms in the 1970s, strictly speaking, should have been conducted at the FAA first and Langley second (where conventional takeoff and landing research took place), and Ames perhaps not at all. Was Ames's management correct to shoehorn the project into the short takeoff and landing program, or should they have shelved the idea because it did not quite fit the Center's prescribed mission? This is no trifle, as air traffic control research at Ames is now a major component of the Agency's aeronautics program.

Budget pressures and government reforms of the last two decades have pushed NASA (and other federal research agencies) to spend more time controlling and monitoring research performance. The irony of this is that there is scant evidence indicating any link between contemporary performance measures and ultimate societal impact.[1] The information that can be gleaned is about project execution. To give an example, the UH-60 Airloads program had its funding cut after a set of lengthy delays. In the interval, between losing funding and actually running out of money, the program attempted to perform as many research flights as possible with their dwindling budget. Interestingly, the limited Airloads data has been a boon to NASA and its partners in industry. Thus, it is entirely possible for a project to have poor performance measures and still have a large, beneficial impact down the road. In the absence of a crystal ball, the trend toward accountability will either have to muddle forward or yield to some realization that innovation is capricious.

1. On this question of linking performance indicators to societal impact, see U.S. General Accounting Office, Report to Congressional Requesters, "Measuring Performance: Strengths and Limitations of Research Indicators" (GAO/RCED-97-91, March 1997).

Why Aeronautics?

With NASA now over 50 years old, aeronautics appears to have survived a decadelong existential crisis. I say "appears" because funding continues on a downward trend and because there is no guarantee that future Administrators, legislatures, and Presidents will value things similarly. Though measures have been taken by both the NASA leadership and the White House to make aeronautics a long-term priority, the fact remains that the Cold War is still over. In the absence of imminent threats to American air superiority, the logic for continued research comes from commerce and the public good. Commercial support, as noted, has become a sensitive topic given trade disputes with the EU over government subsidies. Still, Europe's aviation industry and its research support pose a competitive threat that is likely to prop up political support for NASA's aeronautics programs. Further out on the horizon, China's expanding aviation industry and growing national R&D capability may do the same.

Aeronautics' survival lay in a wealth of unanswered questions and an institutional ability to master new fields. Underlying this has been a durable, grassroots enthusiasm for doing esoteric research and solving puzzles. In spite of top-level vision statements, aeronautics' much smaller undertaking has been largely self-motivated. Taken with the Agency's experimental facilities, the system remains a prized national resource. While aeronautics funding continues to flirt with lower and lower levels of budgetary support, there is likely great reluctance to find out the true cost of doing without the First A.

Appendix:

Aeronautics Budget

Any long-run graph of NASA's aeronautics funding must be viewed with care and skepticism. Budget categories for aeronautics have changed, especially over the last two decades. What NASA defined as aeronautics one year could become something else the next. Accounting methods also changed, the most dramatic being the transition to full-cost accounting in 2004.

Figure 1 shows R&D and total funding (including R&D, facilities construction and maintenance, and research and program management). The R&D figures (e.g., what is allocated to specific research projects) are the most difficult to derive and rely on a number of sources. The R&D line jumps dramatically, of course, with the transition to full-cost accounting, approaching the total allocation for aeronautics. The 2004 President's Budget lists the 2003 Aeronautics Technology budget at $1.006 billion and gives $599 million as the pre-full-cost accounting budget, a 67 percent increase. Figure 2 shows this information in 2008 adjusted dollars.

Figure 3 shows NASA's aeronautics authorization against NASA's total authorization. The data begin at 1962 because aeronautics was not broken out of the Agency's numbers prior to that year. Figure 4 shows the data in 2008 adjusted dollars.

Figure 1. NASA Aeronautics Budget, 1959–2008

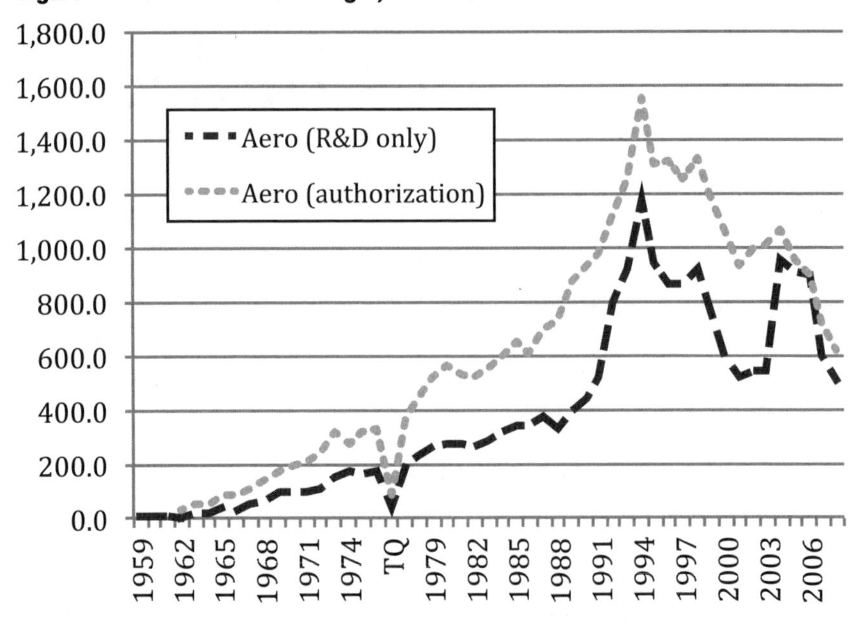

Figure 2. NASA Aeronautics Budget, 1959–2008, FY 2008 Dollars

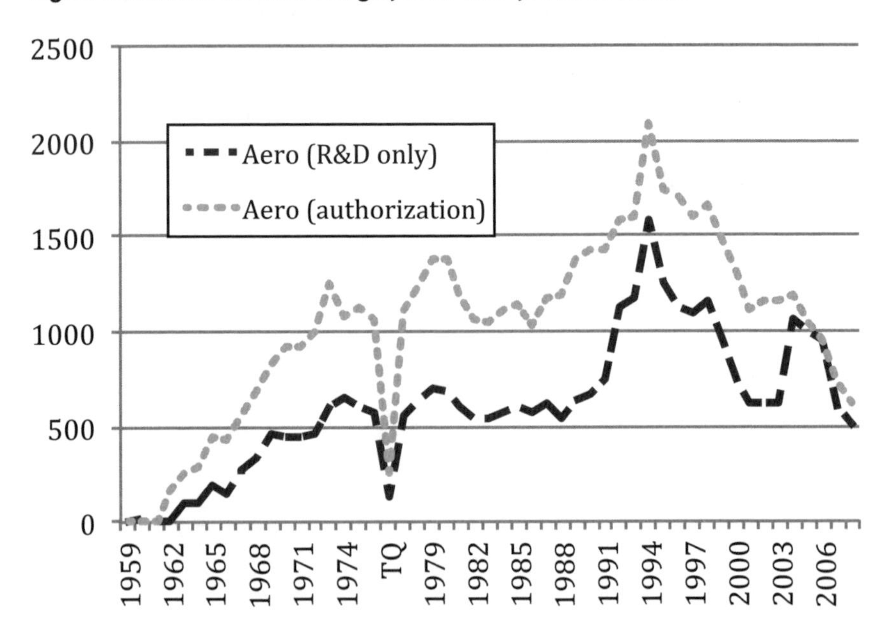

Figure 3. NASA Aeronautics and Agency Budget Authorization, 1962–2008

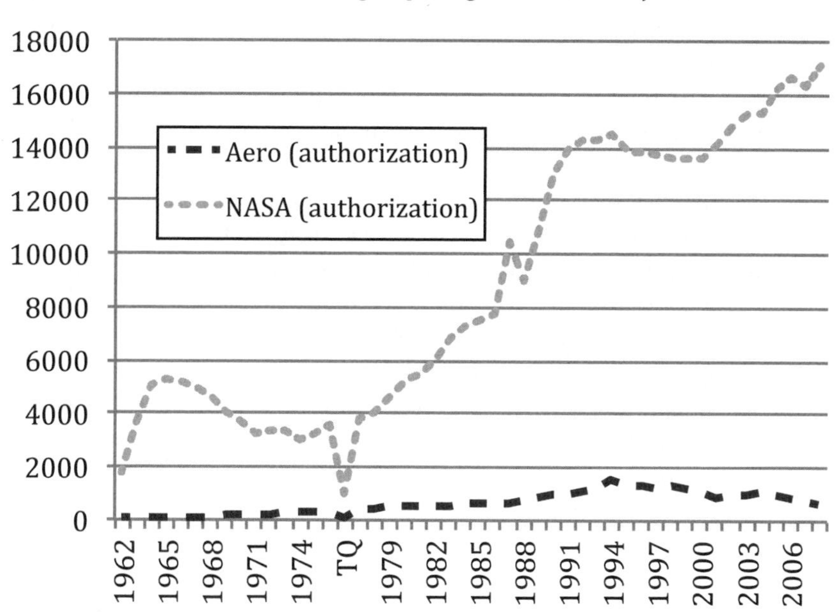

Figure 4. NASA Aeronautics and Agency Budget Authorization, 1962–2008, FY 2008 Dollars

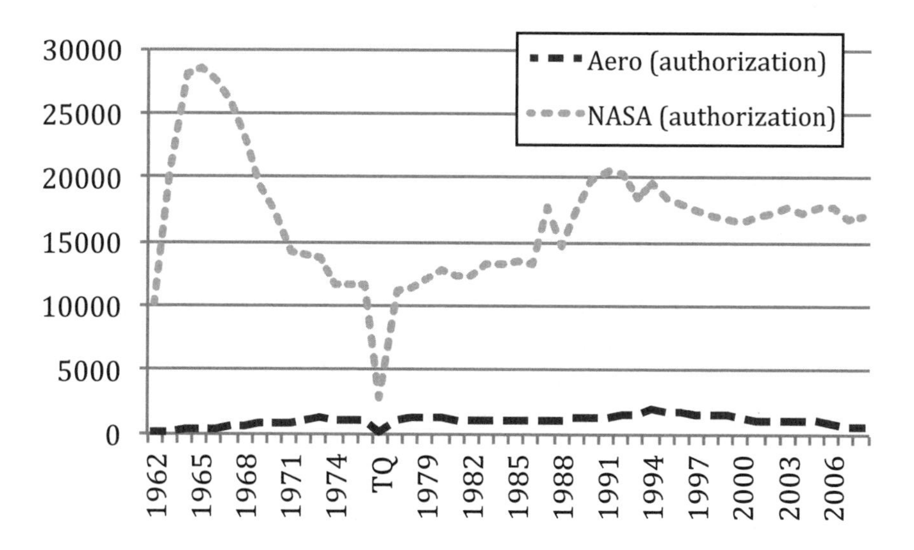

Notes

- NASA's accounting codes were under development from 1959 to 1963.
- In 1976, the fiscal year changed from 1 July–30 June to 1 October–30 September. The three-month "transition quarter" is denoted as "TQ."
- Aeronautics restructuring in the late 1990s and early 2000s makes it difficult to define a consistent R&D budget. To this end, the graph relies significantly on data supplied directly to the author from the Aeronautics Research Mission Directorate.
- The spike in R&D funding in 2004 derives primarily from the change to full-cost accounting.
- The inflation factors for fiscal year 2008 are taken from *Aeronautics and Space Report of the President, Fiscal Year 2008* Activities, Appendix D-1B.

Sources

Total aeronautics expenditures and authorization, as well as Agency expenditures and authorization, derive from the Report to the Congress from the President of the United States, *United States Aeronautics and Space Activities*, later titled *Aeronautics and Space Report of the President*, 1959 to 2008 inclusive. The aeronautics R&D funding data are derived from many sources, including the following:

1959–68

NASA Historical Data Book series (NASA SP-4012): Jane Van Nimmen and Leonard C. Bruno with Robert L. Rosholt, *NASA Historical Data Book*, vol. 1, *NASA Resources 1958–1968* (Washington, DC: NASA SP-4012, 1988), table 4-22.

1969

Data supplied to the author by Cindy Brumfield, Deputy Director, Resources Management Division, Aeronautics Research Mission Directorate. These data included long-run R&D numbers, annotated congressional budget data from 1988 to 1997, and annotated President's budget data from 1998 to 2003.

1970–78

Ihor Gawdiak with Helen Fedor, *NASA Historical Data Book*, vol. 4, *NASA Resources 1969–78* (Washington, DC: NASA SP-4012, 1994), table 4-22.

1979–88

Judy A. Rumerman, *NASA Historical Data Book*, vol. 6, *NASA Space Applications, Aeronautics and Space Research and Technology, Tracking and Data Acquisition/Support Operations, Commercial Programs, and Resources 1979–88* (Washington, DC: NASA SP-4012, 2000), table 8-12.

1989–2003

Data supplied to the author by Cindy Brumfield, Deputy Director, Resources Management Division, Aeronautics Research Mission Directorate. These data included long-run R&D numbers, annotated congressional budget data from 1988 to 1997, and annotated President's budget data from 1998 to 2003.

2004–10

NASA annual budgets and strategic plans.

The NASA History Series

Reference Works, NASA SP-4000:

Grimwood, James M. *Project Mercury: A Chronology*. NASA SP-4001, 1963.

Grimwood, James M., and Barton C. Hacker, with Peter J. Vorzimmer. *Project Gemini Technology and Operations: A Chronology*. NASA SP-4002, 1969.

Link, Mae Mills. *Space Medicine in Project Mercury*. NASA SP-4003, 1965.

Astronautics and Aeronautics, 1963: Chronology of Science, Technology, and Policy. NASA SP-4004, 1964.

Astronautics and Aeronautics, 1964: Chronology of Science, Technology, and Policy. NASA SP-4005, 1965.

Astronautics and Aeronautics, 1965: Chronology of Science, Technology, and Policy. NASA SP-4006, 1966.

Astronautics and Aeronautics, 1966: Chronology of Science, Technology, and Policy. NASA SP-4007, 1967.

Astronautics and Aeronautics, 1967: Chronology of Science, Technology, and Policy. NASA SP-4008, 1968.

Ertel, Ivan D., and Mary Louise Morse. *The Apollo Spacecraft: A Chronology, Volume I, Through November 7, 1962*. NASA SP-4009, 1969.

Morse, Mary Louise, and Jean Kernahan Bays. *The Apollo Spacecraft: A Chronology, Volume II, November 8, 1962–September 30, 1964*. NASA SP-4009, 1973.

Brooks, Courtney G., and Ivan D. Ertel. *The Apollo Spacecraft: A Chronology, Volume III, October 1, 1964–January 20, 1966*. NASA SP-4009, 1973.

Ertel, Ivan D., and Roland W. Newkirk, with Courtney G. Brooks. *The Apollo Spacecraft: A Chronology, Volume IV, January 21, 1966–July 13, 1974*. NASA SP-4009, 1978.

Astronautics and Aeronautics, 1968: Chronology of Science, Technology, and Policy. NASA SP-4010, 1969.

Newkirk, Roland W., and Ivan D. Ertel, with Courtney G. Brooks. *Skylab: A Chronology*. NASA SP-4011, 1977.

Van Nimmen, Jane, and Leonard C. Bruno, with Robert L. Rosholt. *NASA Historical Data Book, Volume I: NASA Resources, 1958–1968*. NASA SP-4012, 1976; rep. ed. 1988.

Ezell, Linda Neuman. *NASA Historical Data Book, Volume II: Programs and Projects, 1958–1968*. NASA SP-4012, 1988.

Ezell, Linda Neuman. *NASA Historical Data Book, Volume III: Programs and Projects, 1969–1978*. NASA SP-4012, 1988.

Gawdiak, Ihor, with Helen Fedor. *NASA Historical Data Book, Volume IV: NASA Resources, 1969–1978*. NASA SP-4012, 1994.

Rumerman, Judy A. *NASA Historical Data Book, Volume V: NASA Launch Systems, Space Transportation, Human Spaceflight, and Space Science, 1979–1988*. NASA SP-4012, 1999.

Rumerman, Judy A. *NASA Historical Data Book, Volume VI: NASA Space Applications, Aeronautics and Space Research and Technology, Tracking and Data Acquisition/Support Operations, Commercial Programs, and Resources, 1979–1988*. NASA SP-4012, 1999.

Rumerman, Judy A. *NASA Historical Data Book, Volume VII: NASA Launch Systems, Space Transportation, Human Spaceflight, and Space Science, 1989–1998*. NASA SP-2009-4012, 2009.

Rumerman, Judy A. *NASA Historical Data Book, Volume VIII: NASA Earth Science and Space Applications, Aeronautics, Technology, and Exploration,*

Tracking and Data Acquisition/Space Operations, Facilities and Resources, 1989–1998. NASA SP-2012-4012, 2012.

No SP-4013.

Astronautics and Aeronautics, 1969: Chronology of Science, Technology, and Policy. NASA SP-4014, 1970.

Astronautics and Aeronautics, 1970: Chronology of Science, Technology, and Policy. NASA SP-4015, 1972.

Astronautics and Aeronautics, 1971: Chronology of Science, Technology, and Policy. NASA SP-4016, 1972.

Astronautics and Aeronautics, 1972: Chronology of Science, Technology, and Policy. NASA SP-4017, 1974.

Astronautics and Aeronautics, 1973: Chronology of Science, Technology, and Policy. NASA SP-4018, 1975.

Astronautics and Aeronautics, 1974: Chronology of Science, Technology, and Policy. NASA SP-4019, 1977.

Astronautics and Aeronautics, 1975: Chronology of Science, Technology, and Policy. NASA SP-4020, 1979.

Astronautics and Aeronautics, 1976: Chronology of Science, Technology, and Policy. NASA SP-4021, 1984.

Astronautics and Aeronautics, 1977: Chronology of Science, Technology, and Policy. NASA SP-4022, 1986.

Astronautics and Aeronautics, 1978: Chronology of Science, Technology, and Policy. NASA SP-4023, 1986.

Astronautics and Aeronautics, 1979–1984: Chronology of Science, Technology, and Policy. NASA SP-4024, 1988.

Astronautics and Aeronautics, 1985: Chronology of Science, Technology, and Policy. NASA SP-4025, 1990.

Noordung, Hermann. *The Problem of Space Travel: The Rocket Motor*. Edited by Ernst Stuhlinger and J. D. Hunley, with Jennifer Garland. NASA SP-4026, 1995.

Gawdiak, Ihor Y., Ramon J. Miro, and Sam Stueland. *Astronautics and Aeronautics, 1986–1990: A Chronology*. NASA SP-4027, 1997.

Gawdiak, Ihor Y., and Charles Shetland. *Astronautics and Aeronautics, 1991–1995: A Chronology*. NASA SP-2000-4028, 2000.

Orloff, Richard W. *Apollo by the Numbers: A Statistical Reference*. NASA SP-2000-4029, 2000.

Lewis, Marieke, and Ryan Swanson. *Astronautics and Aeronautics: A Chronology, 1996–2000*. NASA SP-2009-4030, 2009.

Ivey, William Noel, and Marieke Lewis. *Astronautics and Aeronautics: A Chronology, 2001–2005*. NASA SP-2010-4031, 2010.

Buchalter, Alice R., and William Noel Ivey. *Astronautics and Aeronautics: A Chronology, 2006*. NASA SP-2011-4032, 2010.

Lewis, Marieke. *Astronautics and Aeronautics: A Chronology, 2007*. NASA SP-2011-4033, 2011.

Lewis, Marieke. *Astronautics and Aeronautics: A Chronology, 2008*. NASA SP-2012-4034, 2012.

Lewis, Marieke. *Astronautics and Aeronautics: A Chronology, 2009*. NASA SP-2012-4035, 2012.

Management Histories, NASA SP-4100:

Rosholt, Robert L. *An Administrative History of NASA, 1958–1963*. NASA SP-4101, 1966.

Levine, Arnold S. *Managing NASA in the Apollo Era*. NASA SP-4102, 1982.

Roland, Alex. *Model Research: The National Advisory Committee for Aeronautics, 1915–1958*. NASA SP-4103, 1985.

Fries, Sylvia D. *NASA Engineers and the Age of Apollo*. NASA SP-4104, 1992.

Glennan, T. Keith. *The Birth of NASA: The Diary of T. Keith Glennan*. Edited by J. D. Hunley. NASA SP-4105, 1993.

Seamans, Robert C. *Aiming at Targets: The Autobiography of Robert C. Seamans*. NASA SP-4106, 1996.

Garber, Stephen J., ed. *Looking Backward, Looking Forward: Forty Years of Human Spaceflight Symposium*. NASA SP-2002-4107, 2002.

Mallick, Donald L., with Peter W. Merlin. *The Smell of Kerosene: A Test Pilot's Odyssey*. NASA SP-4108, 2003.

Iliff, Kenneth W., and Curtis L. Peebles. *From Runway to Orbit: Reflections of a NASA Engineer*. NASA SP-2004-4109, 2004.

Chertok, Boris. *Rockets and People, Volume I*. NASA SP-2005-4110, 2005.

Chertok, Boris. *Rockets and People: Creating a Rocket Industry, Volume II*. NASA SP-2006-4110, 2006.

Chertok, Boris. *Rockets and People: Hot Days of the Cold War, Volume III*. NASA SP-2009-4110, 2009.

Chertok, Boris. *Rockets and People: The Moon Race, Volume IV*. NASA SP-2011-4110, 2011.

Laufer, Alexander, Todd Post, and Edward Hoffman. *Shared Voyage: Learning and Unlearning from Remarkable Projects*. NASA SP-2005-4111, 2005.

Dawson, Virginia P., and Mark D. Bowles. *Realizing the Dream of Flight: Biographical Essays in Honor of the Centennial of Flight, 1903–2003*. NASA SP-2005-4112, 2005.

Mudgway, Douglas J. *William H. Pickering: America's Deep Space Pioneer*. NASA SP-2008-4113, 2008.

Wright, Rebecca, Sandra Johnson, and Steven J. Dick. *NASA at 50: Interviews with NASA's Senior Leadership*. NASA SP-2012-4114, 2012.

Project Histories, NASA SP-4200:

Swenson, Loyd S., Jr., James M. Grimwood, and Charles C. Alexander. *This New Ocean: A History of Project Mercury*. NASA SP-4201, 1966; rep. ed. 1999.

Green, Constance McLaughlin, and Milton Lomask. *Vanguard: A History*. NASA SP-4202, 1970; rep. ed. Smithsonian Institution Press, 1971.

Hacker, Barton C., and James M. Grimwood. *On the Shoulders of Titans: A History of Project Gemini*. NASA SP-4203, 1977; rep. ed. 2002.

Benson, Charles D., and William Barnaby Faherty. *Moonport: A History of Apollo Launch Facilities and Operations*. NASA SP-4204, 1978.

Brooks, Courtney G., James M. Grimwood, and Loyd S. Swenson, Jr. *Chariots for Apollo: A History of Manned Lunar Spacecraft*. NASA SP-4205, 1979.

Bilstein, Roger E. *Stages to Saturn: A Technological History of the Apollo/Saturn Launch Vehicles*. NASA SP-4206, 1980 and 1996.

No SP-4207.

Compton, W. David, and Charles D. Benson. *Living and Working in Space: A History of Skylab*. NASA SP-4208, 1983.

Ezell, Edward Clinton, and Linda Neuman Ezell. *The Partnership: A History of the Apollo-Soyuz Test Project*. NASA SP-4209, 1978.

Hall, R. Cargill. *Lunar Impact: A History of Project Ranger*. NASA SP-4210, 1977.

Newell, Homer E. *Beyond the Atmosphere: Early Years of Space Science*. NASA SP-4211, 1980.

Ezell, Edward Clinton, and Linda Neuman Ezell. *On Mars: Exploration of the Red Planet, 1958–1978*. NASA SP-4212, 1984.

Pitts, John A. *The Human Factor: Biomedicine in the Manned Space Program to 1980*. NASA SP-4213, 1985.

Compton, W. David. *Where No Man Has Gone Before: A History of Apollo Lunar Exploration Missions*. NASA SP-4214, 1989.

Naugle, John E. *First Among Equals: The Selection of NASA Space Science Experiments*. NASA SP-4215, 1991.

Wallace, Lane E. *Airborne Trailblazer: Two Decades with NASA Langley's 737 Flying Laboratory*. NASA SP-4216, 1994.

Butrica, Andrew J., ed. *Beyond the Ionosphere: Fifty Years of Satellite Communications*. NASA SP-4217, 1997.

Butrica, Andrew J. *To See the Unseen: A History of Planetary Radar Astronomy*. NASA SP-4218, 1996.

Mack, Pamela E., ed. *From Engineering Science to Big Science: The NACA and NASA Collier Trophy Research Project Winners*. NASA SP-4219, 1998.

Reed, R. Dale. *Wingless Flight: The Lifting Body Story*. NASA SP-4220, 1998.

Heppenheimer, T. A. *The Space Shuttle Decision: NASA's Search for a Reusable Space Vehicle*. NASA SP-4221, 1999.

Hunley, J. D., ed. *Toward Mach 2: The Douglas D-558 Program*. NASA SP-4222, 1999.

Swanson, Glen E., ed. *"Before This Decade Is Out..." Personal Reflections on the Apollo Program*. NASA SP-4223, 1999.

Tomayko, James E. *Computers Take Flight: A History of NASA's Pioneering Digital Fly-By-Wire Project*. NASA SP-4224, 2000.

Morgan, Clay. *Shuttle-Mir: The United States and Russia Share History's Highest Stage*. NASA SP-2001-4225, 2001.

Leary, William M. *"We Freeze to Please": A History of NASA's Icing Research Tunnel and the Quest for Safety*. NASA SP-2002-4226, 2002.

Mudgway, Douglas J. *Uplink-Downlink: A History of the Deep Space Network, 1957–1997*. NASA SP-2001-4227, 2001.

No SP-4228 or SP-4229.

Dawson, Virginia P., and Mark D. Bowles. *Taming Liquid Hydrogen: The Centaur Upper Stage Rocket, 1958–2002*. NASA SP-2004-4230, 2004.

Meltzer, Michael. *Mission to Jupiter: A History of the Galileo Project*. NASA SP-2007-4231, 2007.

Heppenheimer, T. A. *Facing the Heat Barrier: A History of Hypersonics*. NASA SP-2007-4232, 2007.

Tsiao, Sunny. *"Read You Loud and Clear!" The Story of NASA's Spaceflight Tracking and Data Network*. NASA SP-2007-4233, 2007.

Meltzer, Michael. *When Biospheres Collide: A History of NASA's Planetary Protection Programs*. NASA SP-2011-4234, 2011.

Center Histories, NASA SP-4300:

Rosenthal, Alfred. *Venture into Space: Early Years of Goddard Space Flight Center*. NASA SP-4301, 1985.

Hartman, Edwin P. *Adventures in Research: A History of Ames Research Center, 1940–1965*. NASA SP-4302, 1970.

Hallion, Richard P. *On the Frontier: Flight Research at Dryden, 1946–1981*. NASA SP-4303, 1984.

Muenger, Elizabeth A. *Searching the Horizon: A History of Ames Research Center, 1940–1976*. NASA SP-4304, 1985.

Hansen, James R. *Engineer in Charge: A History of the Langley Aeronautical Laboratory, 1917–1958*. NASA SP-4305, 1987.

Dawson, Virginia P. *Engines and Innovation: Lewis Laboratory and American Propulsion Technology*. NASA SP-4306, 1991.

Dethloff, Henry C. *"Suddenly Tomorrow Came…": A History of the Johnson Space Center, 1957–1990*. NASA SP-4307, 1993.

Hansen, James R. *Spaceflight Revolution: NASA Langley Research Center from Sputnik to Apollo*. NASA SP-4308, 1995.

Wallace, Lane E. *Flights of Discovery: An Illustrated History of the Dryden Flight Research Center*. NASA SP-4309, 1996.

Herring, Mack R. *Way Station to Space: A History of the John C. Stennis Space Center*. NASA SP-4310, 1997.

Wallace, Harold D., Jr. *Wallops Station and the Creation of an American Space Program*. NASA SP-4311, 1997.

Wallace, Lane E. *Dreams, Hopes, Realities. NASA's Goddard Space Flight Center: The First Forty Years*. NASA SP-4312, 1999.

Dunar, Andrew J., and Stephen P. Waring. *Power to Explore: A History of Marshall Space Flight Center, 1960–1990*. NASA SP-4313, 1999.

Bugos, Glenn E. *Atmosphere of Freedom: Sixty Years at the NASA Ames Research Center*. NASA SP-2000-4314, 2000.

No SP-4315.

Schultz, James. *Crafting Flight: Aircraft Pioneers and the Contributions of the Men and Women of NASA Langley Research Center*. NASA SP-2003-4316, 2003.

Bowles, Mark D. *Science in Flux: NASA's Nuclear Program at Plum Brook Station, 1955–2005*. NASA SP-2006-4317, 2006.

Wallace, Lane E. *Flights of Discovery: An Illustrated History of the Dryden Flight Research Center*. NASA SP-2007-4318, 2007. Revised version of NASA SP-4309.

Arrighi, Robert S. *Revolutionary Atmosphere: The Story of the Altitude Wind Tunnel and the Space Power Chambers*. NASA SP-2010-4319, 2010.

Bugos, Glenn E. *Atmosphere of Freedom: Seventy Years at the NASA Ames Research Center*. NASA SP-2010-4314, 2010. Revised Version of NASA SP-2000-4314.

General Histories, NASA SP-4400:

Corliss, William R. *NASA Sounding Rockets, 1958–1968: A Historical Summary*. NASA SP-4401, 1971.

Wells, Helen T., Susan H. Whiteley, and Carrie Karegeannes. *Origins of NASA Names*. NASA SP-4402, 1976.

Anderson, Frank W., Jr. *Orders of Magnitude: A History of NACA and NASA, 1915–1980*. NASA SP-4403, 1981.

Sloop, John L. *Liquid Hydrogen as a Propulsion Fuel, 1945–1959*. NASA SP-4404, 1978.

Roland, Alex. *A Spacefaring People: Perspectives on Early Spaceflight*. NASA SP-4405, 1985.

Bilstein, Roger E. *Orders of Magnitude: A History of the NACA and NASA, 1915–1990*. NASA SP-4406, 1989.

Logsdon, John M., ed., with Linda J. Lear, Jannelle Warren Findley, Ray A. Williamson, and Dwayne A. Day. *Exploring the Unknown: Selected Documents in the History of the U.S. Civil Space Program, Volume I: Organizing for Exploration*. NASA SP-4407, 1995.

Logsdon, John M., ed., with Dwayne A. Day and Roger D. Launius. *Exploring the Unknown: Selected Documents in the History of the U.S. Civil Space Program, Volume II: External Relationships*. NASA SP-4407, 1996.

Logsdon, John M., ed., with Roger D. Launius, David H. Onkst, and Stephen J. Garber. *Exploring the Unknown: Selected Documents in the History of the U.S. Civil Space Program, Volume III: Using Space*. NASA SP-4407, 1998.

Logsdon, John M., ed., with Ray A. Williamson, Roger D. Launius, Russell J. Acker, Stephen J. Garber, and Jonathan L. Friedman. *Exploring the Unknown: Selected Documents in the History of the U.S. Civil Space Program, Volume IV: Accessing Space*. NASA SP-4407, 1999.

Logsdon, John M., ed., with Amy Paige Snyder, Roger D. Launius, Stephen J. Garber, and Regan Anne Newport. *Exploring the Unknown: Selected Documents in the History of the U.S. Civil Space Program, Volume V: Exploring the Cosmos*. NASA SP-2001-4407, 2001.

Logsdon, John M., ed., with Stephen J. Garber, Roger D. Launius, and Ray A. Williamson. *Exploring the Unknown: Selected Documents in the History of the U.S. Civil Space Program, Volume VI: Space and Earth Science.* NASA SP-2004-4407, 2004.

Logsdon, John M., ed., with Roger D. Launius. *Exploring the Unknown: Selected Documents in the History of the U.S. Civil Space Program, Volume VII: Human Spaceflight: Projects Mercury, Gemini, and Apollo.* NASA SP-2008-4407, 2008.

Siddiqi, Asif A., *Challenge to Apollo: The Soviet Union and the Space Race, 1945–1974.* NASA SP-2000-4408, 2000.

Hansen, James R., ed. *The Wind and Beyond: Journey into the History of Aerodynamics in America, Volume 1: The Ascent of the Airplane.* NASA SP-2003-4409, 2003.

Hansen, James R., ed. *The Wind and Beyond: Journey into the History of Aerodynamics in America, Volume 2: Reinventing the Airplane.* NASA SP-2007-4409, 2007.

Hogan, Thor. *Mars Wars: The Rise and Fall of the Space Exploration Initiative.* NASA SP-2007-4410, 2007. Vakoch, Douglas A., ed. *Psychology of Space Exploration: Contemporary Research in Historical Perspective.* NASA SP-2011-4411, 2011.

Monographs in Aerospace History, NASA SP-4500:

Launius, Roger D., and Aaron K. Gillette, comps. *Toward a History of the Space Shuttle: An Annotated Bibliography.* Monographs in Aerospace History, No. 1, 1992.

Launius, Roger D., and J. D. Hunley, comps. *An Annotated Bibliography of the Apollo Program.* Monographs in Aerospace History, No. 2, 1994.

Launius, Roger D. *Apollo: A Retrospective Analysis.* Monographs in Aerospace History, No. 3, 1994.

Hansen, James R. *Enchanted Rendezvous: John C. Houbolt and the Genesis of the Lunar-Orbit Rendezvous Concept.* Monographs in Aerospace History, No. 4, 1995.

Gorn, Michael H. *Hugh L. Dryden's Career in Aviation and Space.* Monographs in Aerospace History, No. 5, 1996.

Powers, Sheryll Goecke. *Women in Flight Research at NASA Dryden Flight Research Center from 1946 to 1995*. Monographs in Aerospace History, No. 6, 1997.

Portree, David S. F., and Robert C. Trevino. *Walking to Olympus: An EVA Chronology*. Monographs in Aerospace History, No. 7, 1997.

Logsdon, John M., moderator. *Legislative Origins of the National Aeronautics and Space Act of 1958: Proceedings of an Oral History Workshop*. Monographs in Aerospace History, No. 8, 1998.

Rumerman, Judy A., comp. *U.S. Human Spaceflight: A Record of Achievement, 1961–1998*. Monographs in Aerospace History, No. 9, 1998.

Portree, David S. F. *NASA's Origins and the Dawn of the Space Age*. Monographs in Aerospace History, No. 10, 1998.

Logsdon, John M. *Together in Orbit: The Origins of International Cooperation in the Space Station*. Monographs in Aerospace History, No. 11, 1998.

Phillips, W. Hewitt. *Journey in Aeronautical Research: A Career at NASA Langley Research Center*. Monographs in Aerospace History, No. 12, 1998.

Braslow, Albert L. *A History of Suction-Type Laminar-Flow Control with Emphasis on Flight Research*. Monographs in Aerospace History, No. 13, 1999.

Logsdon, John M., moderator. *Managing the Moon Program: Lessons Learned from Apollo*. Monographs in Aerospace History, No. 14, 1999.

Perminov, V. G. *The Difficult Road to Mars: A Brief History of Mars Exploration in the Soviet Union*. Monographs in Aerospace History, No. 15, 1999.

Tucker, Tom. *Touchdown: The Development of Propulsion Controlled Aircraft at NASA Dryden*. Monographs in Aerospace History, No. 16, 1999.

Maisel, Martin, Demo J. Giulanetti, and Daniel C. Dugan. *The History of the XV-15 Tilt Rotor Research Aircraft: From Concept to Flight*. Monographs in Aerospace History, No. 17, 2000. NASA SP-2000-4517.

Jenkins, Dennis R. *Hypersonics Before the Shuttle: A Concise History of the X-15 Research Airplane*. Monographs in Aerospace History, No. 18, 2000. NASA SP-2000-4518.

Chambers, Joseph R. *Partners in Freedom: Contributions of the Langley Research Center to U.S. Military Aircraft of the 1990s*. Monographs in Aerospace History, No. 19, 2000. NASA SP-2000-4519.

Waltman, Gene L. *Black Magic and Gremlins: Analog Flight Simulations at NASA's Flight Research Center*. Monographs in Aerospace History, No. 20, 2000. NASA SP-2000-4520.

Portree, David S. F. *Humans to Mars: Fifty Years of Mission Planning, 1950–2000*. Monographs in Aerospace History, No. 21, 2001. NASA SP-2001-4521.

Thompson, Milton O., with J. D. Hunley. *Flight Research: Problems Encountered and What They Should Teach Us*. Monographs in Aerospace History, No. 22, 2001. NASA SP-2001-4522.

Tucker, Tom. *The Eclipse Project*. Monographs in Aerospace History, No. 23, 2001. NASA SP-2001-4523.

Siddiqi, Asif A. *Deep Space Chronicle: A Chronology of Deep Space and Planetary Probes, 1958–2000*. Monographs in Aerospace History, No. 24, 2002. NASA SP-2002-4524.

Merlin, Peter W. *Mach 3+: NASA/USAF YF-12 Flight Research, 1969–1979*. Monographs in Aerospace History, No. 25, 2001. NASA SP-2001-4525.

Anderson, Seth B. *Memoirs of an Aeronautical Engineer: Flight Tests at Ames Research Center: 1940–1970*. Monographs in Aerospace History, No. 26, 2002. NASA SP-2002-4526.

Renstrom, Arthur G. *Wilbur and Orville Wright: A Bibliography Commemorating the One-Hundredth Anniversary of the First Powered Flight on December 17, 1903*. Monographs in Aerospace History, No. 27, 2002. NASA SP-2002-4527.

No monograph 28.

Chambers, Joseph R. *Concept to Reality: Contributions of the NASA Langley Research Center to U.S. Civil Aircraft of the 1990s*. Monographs in Aerospace History, No. 29, 2003. NASA SP-2003-4529.

Peebles, Curtis, ed. *The Spoken Word: Recollections of Dryden History, The Early Years*. Monographs in Aerospace History, No. 30, 2003. NASA SP-2003-4530.

Jenkins, Dennis R., Tony Landis, and Jay Miller. *American X-Vehicles: An Inventory—X-1 to X-50*. Monographs in Aerospace History, No. 31, 2003. NASA SP-2003-4531.

Renstrom, Arthur G. *Wilbur and Orville Wright: A Chronology Commemorating the One-Hundredth Anniversary of the First Powered Flight on December 17, 1903*. Monographs in Aerospace History, No. 32, 2003. NASA SP-2003-4532.

Bowles, Mark D., and Robert S. Arrighi. *NASA's Nuclear Frontier: The Plum Brook Research Reactor*. Monographs in Aerospace History, No. 33, 2004. NASA SP-2004-4533.

Wallace, Lane, and Christian Gelzer. *Nose Up: High Angle-of-Attack and Thrust Vectoring Research at NASA Dryden, 1979–2001*. Monographs in Aerospace History, No. 34, 2009. NASA SP-2009-4534.

Matranga, Gene J., C. Wayne Ottinger, Calvin R. Jarvis, and D. Christian Gelzer. *Unconventional, Contrary, and Ugly: The Lunar Landing Research Vehicle*. Monographs in Aerospace History, No. 35, 2006. NASA SP-2004-4535.

McCurdy, Howard E. *Low-Cost Innovation in Spaceflight: The History of the Near Earth Asteroid Rendezvous (NEAR) Mission*. Monographs in Aerospace History, No. 36, 2005. NASA SP-2005-4536.

Seamans, Robert C., Jr. *Project Apollo: The Tough Decisions*. Monographs in Aerospace History, No. 37, 2005. NASA SP-2005-4537.

Lambright, W. Henry. *NASA and the Environment: The Case of Ozone Depletion*. Monographs in Aerospace History, No. 38, 2005. NASA SP-2005-4538.

Chambers, Joseph R. *Innovation in Flight: Research of the NASA Langley Research Center on Revolutionary Advanced Concepts for Aeronautics*. Monographs in Aerospace History, No. 39, 2005. NASA SP-2005-4539.

Phillips, W. Hewitt. *Journey into Space Research: Continuation of a Career at NASA Langley Research Center*. Monographs in Aerospace History, No. 40, 2005. NASA SP-2005-4540.

Rumerman, Judy A., Chris Gamble, and Gabriel Okolski, comps. *U.S. Human Spaceflight: A Record of Achievement, 1961–2006*. Monographs in Aerospace History, No. 41, 2007. NASA SP-2007-4541.

Peebles, Curtis. *The Spoken Word: Recollections of Dryden History Beyond the Sky*. Monographs in Aerospace History, No. 42, 2011. NASA SP-2011-4542.

Dick, Steven J., Stephen J. Garber, and Jane H. Odom. *Research in NASA History*. Monographs in Aerospace History, No. 43, 2009. NASA SP-2009-4543.

Merlin, Peter W. *Ikhana: Unmanned Aircraft System Western States Fire Missions*. Monographs in Aerospace History, No. 44, 2009. NASA SP-2009-4544.

Fisher, Steven C., and Shamim A. Rahman. *Remembering the Giants: Apollo Rocket Propulsion Development*. Monographs in Aerospace History, No. 45, 2009. NASA SP-2009-4545.

Gelzer, Christian. *Fairing Well: From Shoebox to Bat Truck and Beyond, Aerodynamic Truck Research at NASA's Dryden Flight Research Center*. Monographs in Aerospace History, No. 46, 2011. NASA SP-2011-4546.

Arrighi, Robert. *Pursuit of Power: NASA's Propulsion Systems Laboratory No. 1 and 2*. Monographs in Aerospace History, No. 48, 2012. NASA SP-2012-4548.

Goodrich, Malinda K., Alice R. Buchalter, and Patrick M. Miller, comps. *Toward a History of the Space Shuttle: An Annotated Bibliography, Part 2 (1992–2011)*. Monographs in Aerospace History, No. 49, 2012. NASA SP-2012-4549.

Electronic Media, NASA SP-4600:

Remembering Apollo 11: The 30th Anniversary Data Archive CD-ROM. NASA SP-4601, 1999.

Remembering Apollo 11: The 35th Anniversary Data Archive CD-ROM. NASA SP-2004-4601, 2004. This is an update of the 1999 edition.

The Mission Transcript Collection: U.S. Human Spaceflight Missions from Mercury Redstone 3 to Apollo 17. NASA SP-2000-4602, 2001.

Shuttle-Mir: The United States and Russia Share History's Highest Stage. NASA SP-2001-4603, 2002.

U.S. Centennial of Flight Commission Presents Born of Dreams—Inspired by Freedom. NASA SP-2004-4604, 2004.

Of Ashes and Atoms: A Documentary on the NASA Plum Brook Reactor Facility. NASA SP-2005-4605, 2005.

Taming Liquid Hydrogen: The Centaur Upper Stage Rocket Interactive CD-ROM. NASA SP-2004-4606, 2004.

Fueling Space Exploration: The History of NASA's Rocket Engine Test Facility DVD. NASA SP-2005-4607, 2005.

Altitude Wind Tunnel at NASA Glenn Research Center: An Interactive History CD-ROM. NASA SP-2008-4608, 2008.

A Tunnel Through Time: The History of NASA's Altitude Wind Tunnel. NASA SP-2010-4609, 2010.

Conference Proceedings, NASA SP-4700:

Dick, Steven J., and Keith Cowing, eds. *Risk and Exploration: Earth, Sea and the Stars.* NASA SP-2005-4701, 2005.

Dick, Steven J., and Roger D. Launius. *Critical Issues in the History of Spaceflight.* NASA SP-2006-4702, 2006.

Dick, Steven J., ed. *Remembering the Space Age: Proceedings of the 50th Anniversary Conference.* NASA SP-2008-4703, 2008.

Dick, Steven J., ed. *NASA's First 50 Years: Historical Perspectives*. NASA SP-2010-4704, 2010.

Societal Impact, NASA SP-4800:

Dick, Steven J., and Roger D. Launius. *Societal Impact of Spaceflight*. NASA SP-2007-4801, 2007.

Dick, Steven J., and Mark L. Lupisella. *Cosmos and Culture: Cultural Evolution in a Cosmic Context*. NASA SP-2009-4802, 2009.

Index

Numbers in **bold** indicate pages with photos and tables.

A

B

C

D

E

G

H

I

J

K

L

M

N

O

P

Q

R

S

T

U

V

W

X

Y

Z

GPO U.S. GOVERNMENT PRINTING OFFICE: 2013—378-574/00028